The Effects of Air Pollution on Cultural Heritage

John Watt · Johan Tidblad · Vladimir
Kucera · Ron Hamilton
Editors

The Effects of Air Pollution on Cultural Heritage

 Springer

Editors

John Watt
Middlesex University
Hendon, UK
j.watt@mdx.ac.uk

Vladimir Kucera
Swerea KIMAB AB, Stockholm
Sweden
vladimir.kucera@swerea.se

Johan Tidblad
Swerea KIMAB AB, Stockholm
Sweden
johan.tidblad@swerea.se

Ron Hamilton
Middlesex University
Hendon, UK
r.hamilton@mdx.ac.uk

TD
883
.E33
2009

ISBN 978-0-387-84892-1   e-ISBN 978-0-387-84893-8
DOI 10.1007/978-0-387-84893-8

Library of Congress Control Number: 2008936210

Cover illustration: The cover image shows cherubs damaged by both soiling and corrosion. They are on the outside of St Mary Woolnoth, a fine Hawksmoor Church in the City of London. Our thanks to The Revd Andrew Walker for his permission.

Printed on acid-free paper

springer.com

Preface

Managing the risk to our heritage is, of course, an enormously diverse and complex task, reflecting as it does the tremendous variety of history, style, art and culture that is represented. We have many different types of monument, they are made of many different materials, they range in age over centuries and they are located in radically different environments. Air pollution is only one of the risks that threaten this heritage and may frequently not be the most pressing. In addition we have the added complication that weathering occurs naturally and indeed is often felt to contribute to a sense of age and serenity that is fundamental to the way that we value our ancient buildings.

The damage done by air pollution, however, is real, measurable and in many cases obvious. Our industrial development has left us with a legacy of faceless statues and blackened buildings that will take many years to repair and conserve, even when pollution levels are sufficiently reduced to make it sensible to do so. There are important questions to be asked. How much damage has been done and is being done? What is this costing us? How can we be practical in our conservation to prevent unnecessary loss while protecting context and artistic merit? How much value do people actually place on intangibles like the peace of a Gothic Cathedral and how can we account for these very real benefits and others like them (such as the desire to pass on our legacy to our children and grandchildren) in order to help us raise the money to carry out our repairs and maintenance?

The threat posed to cultural heritage, especially built heritage, by air pollution has been studied for many years and this book is designed to bring together a number of strands of that research to make it accessible to the people responsible for looking after our historic buildings, monuments and artefacts. It will help both these heritage managers to prioritise conservation action in response to this threat within the context of other risks and also environmental policy makers to evaluate the economic benefit of taking action to improve air quality.

We look at the way that buildings weather in the natural environment and then show how pollution adds an extra dimension of damage. We focus on two types of damage – corrosion and soiling – and also briefly review an emerging area of research, the role of air pollution in affecting bio-deterioration of buildings. To develop this discussion we need to present the results of a number of scientific studies. First of all we look at current, past and projected levels of the pollutants that cause the damage. The picture has changed dramatically over the years. Before the policy actions to reduce coal burning, pioneered by the Clean Air Act in the UK but now reflected throughout the developed world, the major corrosion was caused by sulphur dioxide (later know as acid rain) and the buildings were darkened by black smoke. We will show how this scene has changed and examine the new, multi-pollutant, urban environment with its lower domestic and industrial emissions but greatly increased traffic. Second, we look at the way that pollution actually attacks buildings and review the findings of a long series of experiments where typical materials have been exposed to a range of different natural and pollution situations across the world. Assessment of the rate at which they are corroded and soiled has allowed scientists to develop equations that predict the amount of damage that will result from a given amount of pollutant. These are known as "dose-response functions" and can be very powerful when we try to assess the harm that might come to a given building in a given environment. Such studies take many years and are therefore very expensive. It is therefore no surprise that dose-response functions are only available for a limited number of materials. We discuss ways to make use of these insights to evaluate pollution impact in any situation. This leads us to the idea that certain materials can be used as indicators for a more general situation and simple test kits produced to utilise them.

This is not just a book about science, however, it is also about geography and economics. Modern map making tools such as geographic information systems are ideal for showing how the risk is distributed spatially. We show how the science discussed above can be mapped – pollution maps are developed into corrosion and soiling maps by application of the dose response functions. One of the themes of this book is scale and maps can provide information at many different scales. This is illustrated in Fig. 1. The risk maps are another way that building managers and owners can access the scientific data. If the risk categories can be made accessible and relevant, then it is relatively simple to locate the particular building or monument on the map and have an estimate of the likely impact.

The damage maps may be developed into cost maps, which illustrate some of the air quality policy implications, if there is good economic data on repair and maintenance costs and on the extent of the material potentially affected (the stock at risk). We discuss a number of studies that have examined these things. The cost estimates are relatively straightforward in area terms (e.g. per square metre of exposed limestone) but it is much more difficult to estimate how much heritage material is affected. We discuss pioneering estimates of what might be termed technical materials (i.e. materials used in houses, factories and infrastructure), which use generalisations about ratios of materials to develop "identikit" buildings whose numbers are then estimated from land use maps or population density. Unfortunately, while it is relatively safe to say that, within a

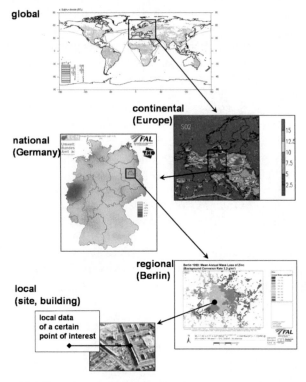

Fig. 1 Maps provide information at many different scales

limited area, most houses are of a certain type, it is certainly not possible to do this for heritage buildings. The latter are, by their nature, less frequent and may reflect a wholly different material makeup due to their importance at time of construction, and therefore use of special materials, or due to their having survived from an earlier period with different construction materials to those used later. We discuss newly emerging research that is starting to address stock at risk inventories for cultural heritage, sometimes including estimates of construction materials.

This is also a book about risk management and policy. We discuss ways that people's values may be brought into decision making. Risk management cannot rely solely on numbers, however much scientists and economists might like it to. Numbers come laden with value judgements anyway, of course, and we discuss the ways that both can inform each other. We show how conservation values such as "truth to original materials" or "reversibility of treatments" can be built into the costs calculations but, just as importantly, we show how it is possible to use peoples' willingness to pay to protect heritage and to develop more equitable business cases for fund raising. We discuss the way economic impact assessments are used in air quality policy making. The cost-benefit analysis in this field rely today largely on human health impact but other costs should also be accounted for, especially impacts on crops, ecosystems and materials. Heritage materials are important here too and people have pointed out that materials may be more sensitive than plants and animals since they have no healing capacity. The final part of our discussion unites all of our threads into an evaluation of what heritage owners and managers can do.

The book has been developed to permit access to the material at a number of different levels. A short overview is presented at the beginning of each chapter to summarise the discussion and place it in the context of the narrative laid out in this preface. Each chapter is a review of the studies undertaken to date within the topic to present the aims and objectives of the research and the main features of the methods used. Results are discussed in terms of the current state of the art and any consensus view that may be articulated. Implications and likely future scenarios are evaluated. These discussions are written for a general reader without assuming prior specialist knowledge and, where technical results are presented, they are fully explained. More specialist readers will find expanded technical detail in the specially created "sources of additional material" sections that close each chapter.

Acknowledgements

A great deal of the research presented in this book was undertaken within a series of projects sponsored by the UNECE and the European Commission and others. The book is the final product of the CULT-STRAT project, also given below, and the European Commission is gratefully acknowledged for the financial support of manuscript preparation and publication costs. The main ones were:

"ICP Materials": The International Co-operative Programme (ICP) on effects on materials including historic and cultural monuments is one of several effect oriented ICPs within the United Nations Economic Commission for Europe (UNECE) and the Convention on Long-range Transboundary Air Pollution (CLRTAP).

"CULT-STRAT" Project: Assessment of Air Pollution Effects on Cultural Heritage – Management Strategies 2004–2007. Contract number: SSPI-CT-2004-501609.

"MULTI-ASSESS" Project: Model for Multi-pollutant Impact and Assessment of Threshold Levels for Cultural Heritage. Contract EVK4-CT-2001-00044 MULTI-ASSESS.

"REACH" Project: Rationalised Economic Appraisal of Cultural Heritage.

EU: Environment and Climate Programme under Topic 2.2.4. PROJECT No: ENV4-CT98-0708 (REACH).

"PPASDC" Project: Particulate Pollution And Stone Damage Contract. EU Contract: EV5V CT94 0519 1/07/94 to 31/10/96.

"EAPMBSP" Project: Effects of Airborne Particulate Matter on Building Surfaces Project. CE Contract STEP-CT90-0097.

We are indebted to our colleagues on all of these projects for their friendship, inspiration and support and to the UNECE including organisations from signatory countries to the Convention on Long-range Transboundary Air Pollution and the European Commission for their financial support. Many of the original contributions to this volume have been made by these colleagues. We are especially grateful to Chrissie Watt for her invaluable assistance in compiling the final manuscript.

We are also grateful to many organisations for their permission to use information which they have made available in the public domain, including: United States Government/USEPA, UK Government/Office of Public Sector Information/DEFRA and the European Environment Agency.

Contents

1 Environment, Pollution and Effects 1
Ron Hamilton and Helen Crabbe

2 Monitoring, Modelling and Mapping 29
Ron Hamilton, Helen Crabbe, Stephan Fitz, and Terje Grøntoft

3 Corrosion 53
Johan Tidblad, Vladimir Kucera, and Susan Sherwood

4 Soiling 105
John Watt, Ron Hamilton, Roger-Alexandre Lefèvre,
and Anda Ionescu

5 Some Aspects of Biological Weathering and Air Pollution 127
Wolfgang Krumbein and Anna Gorbushina

6 Stock at Risk 147
John Watt, Stefan Doytchinov, Roger-Alexandre Lefèvre,
Anda Ionescu, Daniel de la Fuente, Kateřina Kreislová,
and Augusto Screpanti

7 Economic Evaluation 189
John Watt, Ståle Navrud, Zuzana Slížková, and Tim Yates

8 Risk Assessment and Management Strategies at Local Level 215
Tim Yates, Miloš Drdácký, Stanislav Pospíšil, and Terje Grøntoft

9 Air Quality Policy 269
James Irwin, Johan Tidblad, and Vladimir Kucera

Index 297

·

Contributors

Helen Crabbe Centre for Decision Analysis and Risk Management, School of Health and Social Sciences, Middlesex University, The Burroughs, London NW4 4BT, United Kingdom, helencrabbe@yahoo.co.uk

Daniel de la Fuente Materials Engineering, Degradation and Durability National Centre for Metallurgical Research (CENIM/CSIC), Avda. Gregorio del Amo 8, 28040 Madrid, Spain, delafuente@cenim.csic.es

Stefan Doytchinov ENEA – Environmental Department, 301 S.P. Anguillarese, Santa Maria di Galeria, I-00100 Rome, CR Casaccia, Italy, doytchinov@casaccia.enea.it

Miloš Drdácký Institute of Theoretical and Applied Mechanics (ITAM), Prosecká 76, 190 00 Prague, Czech Republic, drdacky@itam.cas.cz

Stephan Fitz Umweltbundesamt, Wörlitzer 1, 06844 Dessau, Germany, stephan.fitz@chello.at

Anna Gorbushina University of Oldenburg, Ammerländer Heerstraße 114-118, D-26129 Oldenburg, Germany, a.gorbushina@uni-oldenburg.de

Terje Grøntoft NILU – Norwegian Institute for Air Research, Urban Environment and Industry, Instituttveien 18, N-2007 Kjeller, Norway, teg@nilu.no

Ron Hamilton Centre for Decision Analysis and Risk Management, School of Health and Social Sciences, Middlesex University, The Burroughs, London NW4 4BT, United Kingdom, r.hamilton@mdx.ac.uk

Anda Ionescu CERTES, University of Paris 12, 61 Avenue du Général de Gaulle, 94010, F-94010 Créteil Cedex, France, ionescu@univ-paris12.fr

James Irwin University of the West of England, Coldharbour Lane, Bristol, BS16 1QY, United Kingdom, jimiirwin@aol.com

Kateřina Kreislová SVUOM, U Mestanskeho pivovaru 934/4, 170 00 Praha 7, Czech Republic, kreislova@svuom.cz

Wolfgang Krumbein University of Oldenburg, Ammerländer Heerstraße 114-118, D-26129 Oldenburg, Germany, wek@uni-oldenburg.de

Vladimir Kucera Swerea KIMAB AB, Box 55970, SE-10216 Stockholm, Sweden, vladimir.kucera@swerea.se

Ståle Navrud Department of Economics and Resource Management, Norwegian University of Life Sciences, 1432 Ås, Norway, stale.navrud@umb.no

Roger-Alexandre Lefèvre Laboratoire Interuniversitaire des Systèmes Atmosphériques, University of Paris 12, 61 Avenue du Général de Gaulle, 94010, F-94010 Créteil Cedex, France, lefevre@lisa.univ-paris12.fr

Stanislav Pospíšil Institute of Theoretical and Applied Mechanics (ITAM), Prosecká 76, 190 00 Prague 9, Czech Republic, pospisil@itam.cas.cz

Augusto Screpanti ENEA–Environmental Department, 301 S.P. Anguillarese, Santa Maria di Galeria, I-00100 Rome, CR Casaccia, Italy

Susan Sherwood Center for Technology and Innovation, P.O. Box 314, Endicott, NY, United States, sisherwood@eartlink.net

Zuzana Slížková Institute of Theoretical and Applied Mechanics (ITAM), Prosecká 76, 190 00 Prague 9, Czech Republic, slizkova@itam.cas.cz

Johan Tidblad Swerea KIMAB AB, Box 55970, SE-10216 Stockholm, Sweden, johan.tidblad@swerea.se

John Watt Centre for Decision Analysis and Risk Management, School of Health and Social Sciences, Middlesex University, The Burroughs, London NW4 4BT, United Kingdom, j.watt@mdx.ac.uk

Tim Yates BRE-Building Research Establishment, Ltd., Garston, Watford WD25 9XX, United Kingdom, Yatest@bre.co.uk

Chapter 1
Environment, Pollution and Effects

Ron Hamilton and Helen Crabbe

1.1 Overview

This chapter will look at the main environmental influences and controls on damage to heritage, which occurs even in the absence of pollution, and also examine the main characteristics and sources of the most important air pollutants that exacerbate this damage or, in some cases, add new types of damage. The types of damage are briefly reviewed at the beginning of the chapter. It is also important to understand the environmental factors that not only influence weathering in the absence of pollution but also are key to the control of pollution damage, and so these are also briefly reviewed.

The danger to heritage from air pollution comes from two main sources – gases that increase the corrosivity of the atmosphere and black particles that dirty light-coloured surfaces. The main mechanism of the former occurs when acid chemicals are incorporated into rain, snow, fog or mist. Familiar as "acid rain", the "acid" comes from oxides of sulphur and nitrogen, largely products of domestic and industrial fuel burning and related to two strong acids: sulphuric acid and nitric acid. Sulphur dioxide (SO_2) and nitrogen oxides (NO_x) released from power stations and other sources form acids where the weather is wet, which fall to the Earth as precipitation and damage both heritage materials and human health. In dry areas, the acid chemicals may become incorporated into dust or smoke, which can deposit on buildings and also cause corrosion when later wetted. Atmospheric chemistry is, of course, far more complex than this and a variety of reactions occur that may form secondary pollutants that also attack materials. One further gas, ozone (O_3), has also been shown to be important. Ozone is a variety of oxygen with three oxygen atoms rather than two as in molecular oxygen. It is the major component of photochemical smog and this ground-level ozone is a product of reactions among the chemicals produced by burning coal, gasoline and other fuels as well as those found in solvents, paints, hairsprays, etc.

R. Hamilton (✉)
Centre for Decision Analysis and Risk Management, School of Health and Social Sciences, Middlesex University, The Burroughs, London NW4 4BT, UK
e-mail: r.hamilton@mdx.ac.uk

J. Watt et al. (eds.), *The Effects of Air Pollution on Cultural Heritage*,
DOI 10.1007/978-0-387-84893-8_1, © Springer Science+Business Media, LLC 2009

Particulate matter is much more complicated because it is a mixture rather than a single substance – it includes dust, soot and other tiny bits of solid materials produced by many sources, including burning of diesel fuel by trucks and buses, incineration of garbage, construction, industrial processes and domestic use of fireplaces and woodstoves. Particulate pollution can cause increased corrosion by involvement in a number of chemical reactions and, often more importantly, it is the source of the black matter that makes buildings dirty.

This chapter looks in more detail at the sources of this pollution, its spatial distribution and trends in emissions over time. As we will see, the picture has changed dramatically over the last fifty years or so, at least in the developed world. The modern urban atmosphere is much less corrosive, in line with major falls in SO_2 brought about in particular by more stringent regulation, but problems remain. Soiling too has changed over time. We will examine the modern emissions pattern, and especially the role of traffic.

1.2 Damage to Cultural Heritage Materials

Managing the risk to our heritage is an enormously diverse and complex task, reflecting as it does a tremendous variety of history, style, art and culture. We have many different types of monuments, made of many different materials, ranging in age over centuries and located in radically different environments. Air pollution is only one of the risks that threaten this heritage and may frequently not be the most pressing. In addition there is the added complication that weathering occurs naturally and indeed is often felt to contribute to a sense of age and serenity that is fundamental to the way we value our ancient buildings. The damage done, however, is real, measurable and in many cases obvious. Historically, industrial development left a legacy of faceless statues and blackened buildings. This was originally seen as a relatively local problem (the damage was caused by emissions from local sources) but the wider scale of the problem was recognised following the acid rain studies in the 1970s.

Buildings weather in the natural environment, but pollution adds an extra dimension of damage (Brimblecombe, 2003; Saiz-Jimenez, 2004). Knowledge of basic damaging mechanisms of historic materials is indispensable for their appropriate and effective protection and safeguarding. In principle, historic materials are deteriorated by means of three mechanisms, which in many cases interact together, simultaneously or in a time sequence.

A very brief description of the different forms of deterioration associated with atmospheric pollution is given below.

Stone
   - Surface erosion and loss of detail
   - Soiling and blackening
   - Biological colonisation
   - Formation of "crust"

Metals
- Surface corrosion
- Development of a stable patina
- Pitting and perforation
- Deterioration/loss of coating (paint, galvanising, etc.)

Timber
- Biological decay
- Deterioration/loss of coating (paint)

Glass
- Corrosion of medieval potash glass
- Soiling of modern soda glass

Other materials
- Concrete
- Mortars
- Brickwork

Structural
- Cracking of walls
- Water penetration

1.2.1 Physical Damage

Material deteriorates mainly mechanically due to the action of external forces (from loads, movements, impacts, human actions, etc.) or internal forces, (e.g. generated by forced deformations at uneven temperature and moisture changes). The time factor may be important for some materials, e.g. long-term overloading of timber. Further, erosion problems decreasing cross-sectional characteristics belong to this group. Physical damage typically results in a mechanical breakdown.

1.2.2 Chemical and Biological Damage

Material is chemically attacked by reactive compounds present in the surrounding environment or produced by biological agents. In the second case we can distinguish between assimilation and dissimilation damage. Where assimilation damage occurs, the material serves as a nutritional source for biological organisms that can, for example, decompose cellulose in organic materials. For dissimilation damage, products released from colonised organisms cause deterioration (biological colonisation can also be protective). Chemical damage can be initiated or accelerated by physical factors, e.g. temperature or light (or another form of radiation), usually termed thermal damage (which is well known for timber and marble) and photochemical damage. Fire damage has a role to play in this category, though in many cases the physical effects of the fire may be dominant. Chemical damage often results in solution and alteration.

1.2.3 Soiling

Soiling is a general phenomenon which decreases the serviceability of historic structures and elements. Defects may be purely aesthetic but more serious failures can occur, e.g. reduction of pipe profiles, bridging of electrical circuits by mycelium, creation of a dark crust causing increased surface heating and elevated temperature stresses. The historic material frequently acts only as a carrier of the undesired layer which might initiate or accelerate the degradation types mentioned above. Soiling can be caused by alterations, deposition and biological colonisation.

1.3 Environmental Factors

It is important to stress that materials degrade even in unpolluted environments. Natural weathering effects are usually added to by pollution and, as we shall see later, this lack of a pollution threshold below which there are no discernable effects has important implications for the establishment of "tolerable" levels of damage and therefore of air quality. Important individual components of weathering are given here:

1.3.1 Radiation

Cultural heritage objects are almost permanently exposed to the action of radiation – whether cosmic, solar or Earth radiation. Physically we distinguish between particulate (nuclear) or electromagnetic radiation (divided according to different wavelengths into radio waves, heat radiation, infrared radiation, light, ultraviolet radiation, x-ray and gamma radiation). Specifically, objects directly exposed to solar radiation exhibit damage and failure due to heat and UV radiation or light. Here the ionic radiation, which is stronger from the point of view of energy, may cause chemical changes in materials (e.g. netting or cleavage of polymers) on one hand, whilst on the other hand the non-ionic radiation exhibits mostly thermal effects.

1.3.2 Temperature

Temperature is one of the influential climatic factors with consequences for cultural heritage damage. Temperature cycles can occur over a very broad band, which is dependent on the geographical position and the time of year.

In some Central Europe and North American locations, temperature falls into a range from about −25°C to about + 30°C. All building materials are sensitive to temperature, and they change their volume in relation to its fluxes. This dilatation is positive (expansion) with temperature increase and negative (shrinkage) with temperature decrease in the majority of building materials. Temperature effects, particularly the temperature gradient, are usually underestimated in practice, even though they are known and published.

Temperature also affects the rate of chemical reactions as well as the time that surfaces remain wet. These factors have implications for the corrosion rates of metals. This is discussed in more detail in Chapter 3.

Temperatures affect emissions of atmospheric pollutants through driving the need for heating and cooling of living space for comfort. This impacts energy use and thus fuel consumption. During cold winter periods, fuel is burnt for space heating, whilst in summer cooling by fans and air conditioning drive energy needs. Increases in extremes in cold and hot periods outside of the normal temperature range as a consequence of climate change will only serve to exacerbate this effect.

1.3.3 Water

Water can affect historic materials and structures in all of its phases and, when acting together with temperature or other factors, can cause deterioration to or even completely destroy a monument. Water acts in its solid phase as ice or snow, and in the fluid phase it attacks as rain, condensation, water trapped in depressions or voids and as underground water which can flow and carry corrosive substances or compounds. The possible erosion of soil under foundations represents another very dangerous phenomenon. In the gaseous state, the water increases the relative humidity of the air and creates conditions for the increase of moisture content of materials as well as providing conditions that encourage biological growths. Wet processes should be avoided or at least substantially reduced to the minimum necessary extent during restoration works.

Among the greater threats to building materials in historic structures are *cyclical changes of moisture content* in materials, namely in the case of a material containing soluble salts (crystallization and hydration pressures generated at reversible changes of phases or crystal phases of salts) or clay materials (swelling of clay). In many cases, a relatively high but stable content of moisture in the building is less harmful than a fluctuating lower moisture content. In some cases, drying of permanently humid materials may be very dangerous (e.g. disintegration of marlstone due to a quick loss of freely bound water).

1.3.4 Air and Air Pollutants

The atmosphere is a thin layer of gas around the Earth. Atmospheric pollutants may be defined as those substances present in the atmosphere that can have adverse effects on health or the environment. The magnitude of the effect is normally closely related to the type and concentration of pollutants to which the systems (humans, ecosystems, buildings, etc.) are exposed. This exposure is a consequence of the location and characteristics of the emitting sources and the prevailing weather conditions.

Gaseous pollutants in the air, mainly oxides of sulphur and nitrogen, have similar effects as the natural atmospheric gases but the effect can be more severe. Gaseous pollution disperses in a shape of a cloud which contains aerosol, salts, acids and alkalis soluble in water. The solid phase particles create an important component of the air pollution, e.g. sand, dust, soot.

Normal gases present in the air (oxygen, ozone and carbon dioxide) react with some materials and may cause their degradation. In recent years the ozone concentration has increased in some locations due to unfavourable factors caused by human impact, and this trend should be stopped if possible. But in many urban areas ozone concentrations are reduced by reactions between the ozone and nitrogen oxide.

1.3.5 Wind

Wind primarily causes loading and mechanical damage of structures; nevertheless, it can also increase or decrease the chemical action of water and gases on cultural heritage objects. The flow around monuments substantially influences the deposition of pollutants, biological colonisation, cycles of drying and wetting, as well as mechanical wear of the attacked surfaces. Wind transports water, salts, dust and gases to the object or building but can also conduct them away in certain conditions.

1.3.6 Meteorological and Climatological Factors

The main constituents of the atmosphere are nitrogen ($N_2 - 78\%$) and oxygen ($O_2 - 21\%$). Three-quarters of the atmosphere's mass is within 15 km of the Earth's surface. Because this thickness is so small compared to the diameter of the Earth (12,742 km) it has often been compared to the skin on an apple. This region of the atmosphere is called the troposphere.

The region between 15 km and 50 km above the Earth's surface is known as the stratosphere. This is the region which contains the ozone layer, which protects the Earth and human health from the sun's ultraviolet radiation. Ninety per cent of all atmospheric ozone is contained within this region. There are very limited mechanisms for air mass movements between the stratosphere and the troposphere but it is possible for some gases that do not experience chemical reactions in the troposphere to disperse into the stratosphere. There is a major concern that chlorine and bromine compounds can find their way into the stratosphere by this route. Chlorofluorocarbons (CFCs or freons) and bromofluorocarbons (halons) are the main sources of chlorine and bromine, and they are used in aerosol propellants, refrigerator coolants and air conditioning units. These compounds react chemically with ozone and lead to a reduction in ozone concentration – leading to what is often referred to as the hole in the ozone layer. Supersonic aircraft fly in the lower levels of the stratosphere and their emissions are thought to be involved in chemical reactions which contribute to stratospheric ozone completion.

Within the troposphere, most air mass movements occur at low altitudes, up to about 2 km. This is known as the mixing layer of the atmosphere. Basically, the Earth's surface is normally warmer than the atmosphere above it, with the result that air is warmer, and consequently less dense, at the earth's surface than at altitude and so turbulence sets in. The degree of turbulence within the mixing layer is a function of temperature, wind speed and surface roughness. Pollutants emitted into this turbulent atmosphere are dispersed and diluted.

It is possible for exceptional circumstances to occur whereby the air at ground level becomes cooler than the air above it. This is known as a temperature inversion. If this happens, and at the same time wind speeds are very low, air mass movements are reduced or even stopped and under these conditions any pollution emitted into the atmosphere will be trapped there. If emissions continue, concentrations will continue to increase. These conditions are known as smog.

Under normal conditions, pollutants emitted to the atmosphere experience dispersion and dilution. The extent of the dispersion depends on the prevailing meteorological factors. The concentrations of the pollutants in the ambient atmosphere is determined by this dispersion, and also by the characteristics of the emitting source(s) – location, height above the ground, speed and temperature of the emissions. The extent of dispersion and dilution within an urban street canyon is very different from that which exists in open countryside. Also, chemical reactions within the atmosphere can change the composition and chemical form of the pollution. These effects can be brought together into a mathematical model (a dispersion model) which allows prediction of pollution concentration, given an input of pollutant emissions and meteorological conditions.

The spatial extent of the dispersion and dilution means that the effects of the emissions may be experienced at local, regional, national or global scales. And equally, the effects experienced at a specific location may be the result of local emissions or emissions a significant distance away from the source. This applies to the full range of adverse effects associated with pollution: from stratospheric ozone depletion due to the use of certain aerosol propellants in homes, to ecosystem damage due to the emission of acidifying compounds in a neighbouring country.

1.3.7 Synergy of Weather Factors

Weather factors mostly act on monuments in a synergistic way. The simultaneous action of temperature and water in repeated freezing/thawing cycles is a typical example of a situation which is very dangerous for wet porous, brittle and quasi-brittle materials. The interaction of temperature and moisture causes repeated and uneven volumetric changes and results in material deterioration and propagation of defects such as cracks. In combination with abrasive particles, the wind can cause remarkable surface erosion (e.g. on monuments in sandy deserts). However, there are numerous other examples: moisture and deposition mechanisms, wind plus water plus pollutants forming weak acids and penetrating into materials.

In recent years it has become clear that there may be one more climatic parameter that must be considered, that is the long-term fluctuations in climate included the long-term trends which are generally grouped together as the "greenhouse effect". A method of examining responses of building and materials to future climates is to review information on the sensitivity of materials to the present climate. However, this shows that there is a substantial lack of understanding and information. Some of the problems stem from the ever-present problem of linking accelerated laboratory testing to performance under atmospheric conditions. Consequently, it is impossible to quantify the impact of a temperature rise of a few degrees, a 10% increase in UV-dose, a 10% change in driving rain, etc. It is said that the climate of a city such as Amsterdam will change into the present climate of Paris but at present it would be difficult to indicate differences in the durability of materials in these two cities.

The effect of climate on construction can be tackled from two directions and both seem to have roles in future research. The first is usually termed "top down" and concentrates on making inventories of failures that occur (or have occurred) in the building stock. Analysis of this type of data can identify the parameters that seem to be associated with failure – for example rain penetration of building fabric – and also indicate how design guidance can be amended to overcome potential problems. Naturally, this type of approach is often closely linked to other research topics, in particular those associated with designing for durability and predicted service life.

The second approach is termed "bottom up" and concentrates on the analysis of mechanisms by which agents (wind, rain, etc.) give rise to failure. At first this approach can seem likely to give more generally applicable results as it is more theoretically based and may well involve modelling of some form. However, experience in other fields, including the climate change modelling, has shown that the width and complexity of the models, combined with many parameter uncertainties, can lead to substantial error bars on the final results. The "top down" approach tends to be more empirical and, although less widely applicable, may give more meaningful and useful results.

The EU-funded "Noah's Ark Project" (ISAC, 2004) assessed a number of factors associated with future climate change which are likely to have an impact on building deterioration at different temporal and spatial scales. It is accepted that trends in atmospheric composition and microclimate parameters will affect the fundamental processes causing damage to building materials, and that increasing atmospheric concentrations of carbon dioxide and other trace gases alter air chemistry and influence chemical reactions. In general, higher temperatures will influence chemical reactions, as well as reduce the effects of thaw-freezing cycles in many environments, while regions with stable cold weather could experience increased thaw-freezing cycles. Alteration in seasonal and annual rainfall will change the length of time during which surfaces are wet, affecting surface leaching and the moisture balance that influences material decay processes.

Although legislation has greatly reduced concentrations of "traditional" air pollutants in cities, the changing climate may enhance the effects of some of them. In particular, changes in wetting and drying cycles on building surfaces may alter the deposition of acidic gases onto the surface. Longer sunlight hours can increase the photochemical degradation of polymers used in both modern construction and the restoration of ancient buildings.

The explanation of damage to cultural heritage buildings from air pollution requires an understanding of many processes, from the details of the building to the nature of the sources of pollution which impact it. The range of interactions is shown diagrammatically in Fig. 1.1. These issues and procedures are developed later in this book.

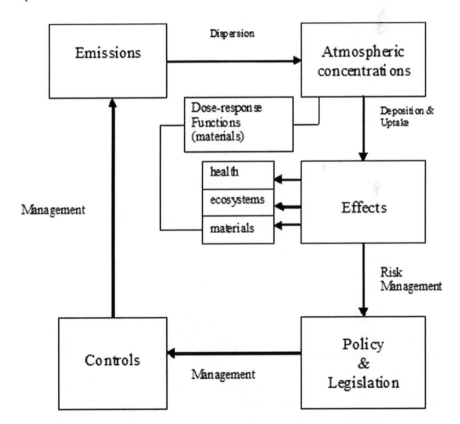

Fig. 1.1 Processes involved in the damage caused by exposure to air pollution

1.4 Pollutant Characteristics

Three broad but useful classifications of atmospheric pollutants are given here.

1.4.1 Natural or Anthropogenic

"Pollutants" from natural sources include lightning, volcanic eruptions, natural microbiological processes in the soil (bacterial action), wind-blown soils and animals. Pollutants from anthropogenic sources, i.e. from man-made activities such as energy generation, transport and domestic functions, include sulphur oxides, nitrogen oxides, hydrocarbons and heavy metals. Most atmospheric pollutants have both natural and anthropogenic sources. However, management and control options for natural source emissions are limited so most emphasis is placed on the identification and control of anthropogenic emissions.

1.4.2 Primary or Secondary

Primary pollutants are those emitted directly to the atmosphere by natural or anthropogenic sources. Secondary pollutants are those produced in the atmosphere by chemical reactions between the primary pollutants, generally in the presence of sunlight. Ozone (O_3) is an example of a secondary pollutant.

1.4.3 Gaseous or Particulate

Some air pollutants exist in the gaseous phase at ambient temperatures, e.g. NO_2, SO_2 and O_3. The normal definition of particulate includes both the liquid and solid forms of matter. Airborne particulate matter includes any material that can be transported through the atmosphere by wind and air movements including particulate matter up to 1 mm in diameter, though most particles in the ambient atmosphere are significantly smaller than this.

   Here we focus on the most important transboundary pollutants that cause acidification (and therefore damage to materials), but also eutrophication and ground-level ozone. These are in turn sulphur dioxide (SO_2), nitrogen dioxide (NO_2), particulate matter (e.g. PM_{10}) and precursor pollutants that cause ground-level ozone (O_3).

1.5 Emission Inventories

In order to calculate changes in emissions to see if international and national emission agreements have been met, government bodies regularly compile and update databases of emissions, called emission inventories. These inventories give emission totals per year for atmospheric pollutants, a breakdown of source contributions and often projections into the future based on the impact of current or future policies. Emission rates are often mapped showing how rates vary on a grid basis or per administration area, country or state. Sources of pollutants are categorised into three types: point sources (those occurring at a particular point, e.g. from industry), line or mobile sources (from transportation) and area sources (from diffuse emission points occurring across a wider area, e.g. agricultural or space heating emissions). Geographical information systems (GIS) are often used to compile the databases of emission sources and rates on a spatial scale and allow for easy mapping or manipulation of the data to show changes over time or compare source contributions.

Estimates of UK emissions are calculated annually on a 1×1 km square grid scale as the National Atmospheric Emissions Inventory (NAEI), by AEA Technology for the UK's Department for Environment, Food and Rural Affairs (DEFRA). Changes in annual emissions are predicted into the future, and also given for each year back to 1970 to show patterns over time. The European Environment Agency's Topic Centre on Air Pollution compiles an emission inventory on a 50×50 km square grid scale. The European Monitoring and Evaluation Programme (EMEP-Co-operative Programme for Monitoring and Evaluation of the Long-range Transmission of Air pollutants in Europe, formerly the CORINAIR inventory) maps emissions across Europe from ten internationally agreed source categories (UN/ECE SNAP nomenclature/source sectors).

The United States Environment Protection Agency (USEPA) National Emission Inventory (NEI) 2002 database includes four of the six criteria pollutants (CO, NO_x, SO_2 and particulate matter (PM_{10} and $PM_{2.5}$)). The NEI also includes emissions of volatile organic compounds (VOCs), which are ozone precursors, emitted from motor vehicle fuel distribution and chemical manufacturing, as well as other solvent uses. VOCs react with nitrogen oxides in the atmosphere to form ozone. Ammonia (NH_3) is also an additional pollutant included in the NEI.

Emissions inventories at a finer spatial scale (0.5 or 1 km square) have been compiled for urban areas and major conurbations. For example, emissions per 1×1 km square grid have been compiled for London since 1993, initially as the London Energy Study, giving emissions of CO_2 and SO_2. Now greatly improved and expanded to 27 pollutants, the London Atmospheric Emissions Inventory (LAEI) compiled by the Greater London Authority (GLA) gives emissions for a base year of 2004. Emission inventory summary data is available to the public and has been used here to examine the main air pollutants and consideration of their sources and emission rates.

1.6 The Main Air Pollutants

This section reviews the main air pollutants responsible for damage to materials, with a summary of their physical and chemical properties. Emission rates are presented. Data for the UK are presented as examples; similar trends are experienced by many other countries.

1.6.1 Sulphur Dioxide

The most important natural sources of sulphur dioxide, SO_2 (and of other sulphur compounds) are volcanoes, during both active and dormant periods. Globally, these contribute perhaps 20% of the world's total sulphur emissions. However, in both developed and less-developed countries, particularly in urban areas, emissions that arise from the combustion of solid fossil fuels are of prime concern. Coal and oil both contain sulphur in varying amounts, and both therefore produce sulphur dioxide when burnt. The average sulphur content of British coal is 1.7%, and that of heavy fuel oil about 2–3%. By contrast, natural gas contains negligible amounts of sulphur.

A summary of the sulphur dioxide yearly emissions since 1970 in the UK is shown in Fig. 1.2

Since 1970 there has been a substantial overall reduction of more than 74% in SO_2 emissions. Reductions in public power, commercial, residential and industrial emissions, changes in fuel use and increasing use of abatement equipment have led to this fall. Emissions of SO_2 from solid fuel use have declined by 84% since 1970 and those from petroleum by 95%. However, coal combustion is still the main source – it accounted for 76% of the 2003 UK SO_2 emissions, for example. The most important factors in the fall in emissions from petroleum use are the decline in fuel oil use and the reduction in the sulphur content of gas oil and diesel fuel. The reduction in the sulphur content of gas oil is particularly significant in sectors such as domestic heating, commercial heating and off-road sources where gas oil is used extensively. The emission profile exhibits a steady decline between 1970 and 2004 with the exception of small peaks in 1973 and 1979 corresponding to the harsh winters in those years and a short period at the end of the 1980s when emissions were relatively constant from year to year. The two main contributors are solid fuel and petroleum products.

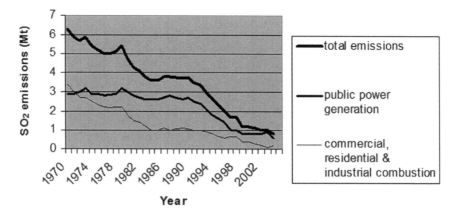

Fig. 1.2 Annual emissions of SO_2 in the UK. (NAEI, 2008)

The effect on ground-level concentrations has been compounded by the change in fuel use patterns that has been going on since the Clean Air Acts of 1956 and 1968. Not only has the burning of coal and oil (once widespread in domestic, commercial and industrial space heating and process use) declined sharply, but electricity generation has been concentrated in large power stations in predominantly rural areas. As a consequence, emissions from low and medium level (in terms of stack height) sources have been replaced by high-level emissions. Higher SO_2 concentrations can therefore be found near coal-fired power stations. There are still a few towns in the UK where the burning of coal for domestic heating produces high concentrations of SO_2 locally, e.g. Belfast City.

It seems likely that national emissions will decline still further as a result of the continued rollout of flue gas desulphurisation in power stations and the replacement of coal by natural gas and imported coal for electricity generation. Relatively small, local sources can have a disproportionate effect on ground-

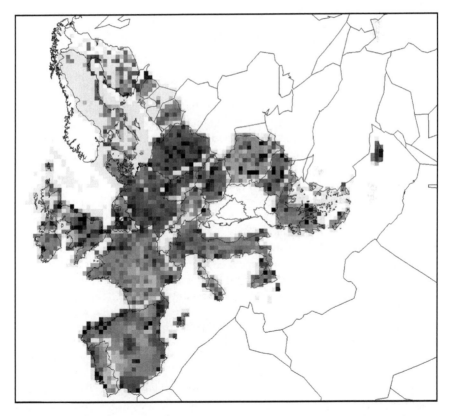

Fig. 1.3 SO_x Emissions per 50 km square grid for the 27 EU countries in 2004 in mega grammes. (EMEP, 2006)

level concentrations, particularly during a high-pollution episode, which occurs as a result of an inversion trapping the emissions from low-level sources within a shallow mixing layer. SO_2 emissions contributed to a number of severe smog pollution episodes in the 1950s and 1960s, for example, the Great Smog in London in December 1952 where industrial and domestic coal burning during a period of cold temperatures and an inversion layer trapped emissions at low levels. Very poor visibility and high levels of SO_2 (up to 3.7 mg m^{-3}) and black smoke (above 4.5 mg m^{-3}) led to up to 4000 extra deaths from respiratory diseases (Anderson, 1999). This was a major drive for government action and helped shape modern air pollution legislation in force today.

European emissions (Fig. 1.3) reflect areas of power use and production. Cities in Spain, for example, can clearly be seen. SO_2 emissions are higher generally in Eastern European countries, due to coal-fired power stations and the relatively high sulphur content of coal sourced locally. This has implications for cultural heritage in cities like Prague for example, as historic buildings are exposed to higher SO_2 concentrations.

1.6.2 Nitrogen Oxides and Nitric Acid

There are a number of nitrogen oxides (NO_x), but the one of principal interest as an air pollutant likely to have adverse effects on human health and soiling properties is nitrogen dioxide (NO_2). Other nitrogen oxides such as N_2O_3 and N_2O_4 are of lesser significance as air pollutants. Nitrous oxide (N_2O) is of some interest as a greenhouse gas, while nitrogen pentoxide (N_2O_5) and nitrogen trioxide (NO_3) are important in the atmospheric chemistry of pollution episodes. Nitrogen compounds are also contributors to the wet and dry deposition of acidic compounds on vegetation and buildings.

The oxide emitted to the atmosphere in the largest amounts is nitric oxide (NO), which is itself relatively innocuous. However, it is readily oxidised – by, for example, ozone – and is the immediate source of most atmospheric nitrogen dioxide. The fraction of nitric oxide that is converted to nitrogen dioxide by ozone and the speed of the reaction depend upon the concentrations of the two reacting gases. The reaction is fast and, provided the concentration of ozone is sufficient, is substantially complete within a minute or so. In locations such as busy roads where the nitric oxide concentration is high, the local concentration of ozone may be insufficient to complete the reaction. Much of the nitric oxide may therefore travel downwind of its source and reach relatively clean air before more ozone becomes available. The NO/NO_2 ratio is therefore significantly larger in urban than in rural areas.

Nitric oxide and nitrogen dioxide are known collectively as NO_x, and it is usual to express NO_x as the NO_2 equivalent in estimates of emissions by mass. In the atmosphere, nitric acid is present in very minute quantities as a gas or vapour, produced by reactions between atmospheric oxidants and emitted nitrogen oxides. Typically, emitted NO_x is converted to nitric acid within a day or less. The atmospheric removal processes for gaseous nitric acid are by wet and dry deposition. The estimated half-life and lifetime for dry deposition of nitric acid are 1.5–2 days and 2–3 days respectively, and there is also efficient nitric acid removal during episodic precipitation events. Nitric acid reacts with gaseous ammonia to form particulate or aerosol nitrate, which in turn is removed by wet and dry deposition of the particles. The average half-life and lifetime for particles in the atmosphere are about 3.5–10 days and 5–15 days respectively.

1.6.2.1 Sources of Atmospheric NO_x

The most important natural sources of NO_x are volcanoes, lightning and bacterial action (natural microbiological processes in the soil) and on a global scale these far outweigh anthropogenic sources. However, in urban areas the emissions arising from the combustion of fossil fuels are of most concern.

Any combustion process may produce nitrogen oxides from the oxidation of nitrogen in the air or, less importantly, in the fuel or other material being burned. The most important man-made sources are the combustion of fossil fuels in power generation and the combustion of petrol and diesel in vehicles. Other relatively minor contributions to the atmosphere come from non-combustion industrial processes such as the manufacture of nitric acid, the use of explosives and welding operations.

Estimates of UK NO_x emissions during the period 1970–2004 and source apportionment are given in Fig. 1.4. It can be seen that the main sources are power stations and road transport; the latter and coal combustion combined to account for 54% of UK emissions in 2003. Emissions from power stations have remained fairly constant during the period, only decreasing from the end of the 1990s from improvements in abatement technology. Those from road transport steadily increased from about 0.6 Mt in 1970 to 1.3 Mt per year in the early

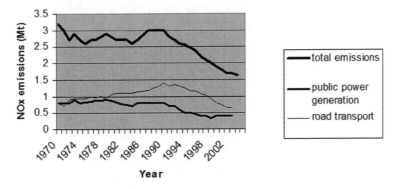

Fig. 1.4 Annual UK NOx emissions and proportion from sources. (NAEI, 2008)

1990s, when they constituted about 50% of the total. Road transport emissions then decreased to around 0.6 t/y today as emissions from petrol engines have declined since the introduction of catalytic converters in the early 1990s. EC Directives on exhaust emissions required mandatory fitting of a catalytic converter (a three-way catalyst, TWC) to all new petrol engine cars throughout the European Union from 1993. NO_x emissions from such cars (expressed in terms of grams per kilometre averaged over the whole TWC car population) are generally at least 10 times lower than from cars without converters. As almost all of the petrol vehicle fleet now has TWCs, the maximum pollution benefit of this technology has been observed and only a relatively small reduction in NO_x emissions will occur in the future as older cars are taken out of service. A 75% reduction in NO_x emissions resulted from this action.

Recent research has highlighted the increasing importance of directly emitted NO_2. There is evidence for significant amounts of NO_2 being emitted directly from the tailpipes of diesel vehicles, especially when slow moving, with levels possibly as high as 25% of total NO_x emissions in mass terms. These primary emissions have a significant impact on roadside NO_2 concentrations in areas where the proportion of diesel vehicles is large (DEFRA, 2004).

Fuel use and population density also drive the spatial pattern of NO_x emissions as Fig. 1.5 shows for emission densities across the United States. Similar patterns occur for all other industrialised nations.

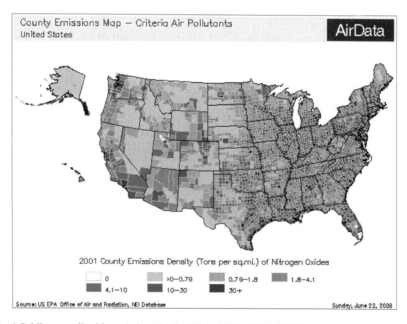

Fig. 1.5 Nitrogen dioxide emissions in the US in 1999. (EPA, 2008)

1.6.3 Particulate Matter

Particulate matter is a term that represents a wide range of chemically and physically diverse substances that can be described by size, formation mechanism, origin, chemical composition, atmospheric behaviour and method of measurement. The concentration of particles in the atmosphere varies across space and time and as a function of the source of the particles and the transformations that occur to them as they age and travel.

There are a number of ways of measuring the amount of PM in the air, which has resulted in a number of different ways of expressing the data. Historically, measurements focused on a weight measure, total suspended particulate (TSP), which was often averaged over seven days, or measured "smoke" as a proxy for particulate. The levels of smoke or black smoke in the atmosphere were measured as an indicator of the concentration of particles in the air. In the UK a measurement of black smoke has been the traditional technique as an indication of the level of PM in the air and is still used in the Automatic Urban Network (AURN) and EU Directive monitoring networks. The method consists of drawing air through a white filter paper and measuring the darkness of the stain produced by use of a reflectometer. A standard calibration curve is used to convert the reflectometer measurement into a nominal mass concentration of the airborne particles, referred to as the "black smoke concentration". The method is defined by British Standards (BSI, 1969). The validity of the results depends upon the fraction of carbonaceous (i.e. black) material in the sample. The standard calibration curve is valid only for particulate material of the type that existed in UK urban areas up to the early 1960s, and nowadays there is no reliable relationship between black smoke and gravimetric measurements (Bailey and Clayton, 1982). The relationship between TSP and PM_{10} is controlled by the size distribution of the airborne particles, and is therefore very variable.

In the last 15–20 years, studies have concentrated on different size fractions due to the nature, sources, behaviour and effects of the PM. The predominant particle mix in most cities is dominated by fine particles (less than 2.5 μm in aerodynamic diameter (see definitions later in this section), known as $PM_{2.5}$) generated by combustion sources, with smaller amounts of coarse dust (between 2.5 and 10 μm in diameter). Particles less than 10 μm in diameter (PM_{10}) are often measured that include both fine and coarse dust particles. More precisely, PM_{10} has been specified in terms of a sampling method for which the collection efficiency of particles of size 10 μm is 50%. These particles pose the greatest health concern because they can pass through the nose and throat and get into the lungs. Particles larger than 10 μm in diameter that are suspended in the air are referred to as total suspended particulates (TSP). A number of size-selective inlets have been designed for the TSP monitors to capture particles, with a cut point of 10 μm.

PM_{10} has become the generally accepted measure of particulate material in the atmosphere in the UK and in Europe due to its well-documented association with adverse health effects. The main sources of primary PM_{10} are road transport

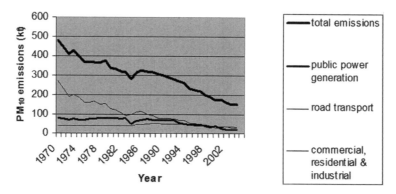

Fig. 1.6 Annual PM_{10} emissions since 1970 in the UK. (NAEI, 2008)

(all road transport emits PM_{10}, but diesel vehicles emit a greater mass of particulate per vehicle kilometre), stationary combustion (domestic coal combustion traditionally being the major source of particulate emissions in the UK) and industrial processes (including bulk handling, construction, mining and quarrying).

UK emissions of PM_{10} declined by 51% between 1990 and 2003, giving an emission total of 0.14 Mt in 2003 (Fig. 1.6). This reflects a trend away from coal use particularly by domestic users. Coal combustion and road transport together contributed 57% of UK emissions of PM_{10} in 2003. PM_{10} emissions from road transport have shown a steady decline across recent years. $PM_{2.5}$ emissions have also fallen, but by a smaller amount, with the largest source sector being road transport, accounting for 52% of the 2003 total emission. Black smoke emissions in the UK have significantly declined (by some 66% between 1970 and 2003). $PM_{2.5}$ emissions in 2003 were estimated to be 152 kt although these estimates are based on old measurement data and are hence very uncertain. Domestic and commercial emissions have fallen from 263 kt (54% of the total emission) in 1970 to 41 kt (27%) in 2004 (NAEI, 2008).

Many of the atmospheric pollutants about which there is concern as to possible adverse health effects and of soiling are in particulate form. Lead and other heavy metals, asbestos and some of the polyaromatic hydrocarbons are well-known examples. The particulate material in the atmosphere consists of many chemically and physically diverse substances that exist as discrete particles (liquid droplets or solids) over a wide range of sizes. They originate from a variety of sources – either emitted directly from, for example, vehicles and factories or formed in the atmosphere by the transformations of gaseous pollutants such as sulphur oxides, nitrogen oxides and volatile organic substances.

Attempts have been made in the past to classify disperse aerosol systems in gases on the basis of their nature, origin and particle size. None of these has been completely successful because of the indefinite character of many types of particulate clouds and the difference, often not clear cut, between scientific and popular descriptive terms (Friedlander, 2000). The precise meaning of terms such as grit, dust, fume, smoke, coarse, fine, inhalable and respirable is not so universally agreed as to absolve the user from giving a definition.

The diameter of an atmospheric particle is the single most influential parameter in determining its aerodynamic properties, including sedimentation, diffusion and impaction. Due to its importance, the parameter *aerodynamic diameter* has been defined. The aerodynamic diameter of a particle is the diameter of a sphere of unit density (1×10^3 kg m^{-3}) having the same terminal settling velocity as the particle in question. For particles of aerodynamic diameter less than 0.5 μm, the particle diffusion diameter should be used instead of the particle aerodynamic diameter. The diffusion diameter of a particle is the diameter of a sphere with the same diffusion coefficient as the particle in question. Unless otherwise stated, the term diameter here means aerodynamic diameter or diffusion diameter as appropriate. The terminal settling velocities of particles of various diameters are given in Table 1.1. These are calculated values; the method of calculation may be found in textbooks such as Hinds (1999) or Friedlander (2000).

In spite of these difficulties, there are some broad patterns as regards origin and size, which are helpful when considering the sources, nature and effects of airborne particulate material. Groups of particles (sometimes called "modes") can often be distinguished.

The **nucleation** mode, the group with the smallest particle size, consists of ions and nuclei (often of the dimensions of molecular clusters), and the particles into which they grow as a consequence of the condensation of vapours upon them. Particles arising from gas to particle conversion (e.g. sulphuric acid droplets from the oxidation of sulphur dioxide) are initially formed by condensation onto a nucleus. The size range of particles in this mode extends from that of molecular clusters – say, 0.001 μm in diameter – to about 0.1 μm. Condensation nuclei are usually present in very large number concentrations in urban atmospheres, but because of their small size they make a relatively small contribution to the total mass concentration.

The **accumulation** mode consists of particles which have grown from the nucleation mode by further condensation of vapours upon them or by coagulation. Their size range is usually taken to be from about 0.1 μm to about 1 μm. They are relatively stable, in that the processes which remove particles from the atmosphere (e.g. diffusion, washout and sedimentation) are least efficient for particles in this size range.

The **coarse** mode consists of particles greater than about 1 μm in diameter. Many of them originate from mechanical processes such as erosion, resuspension and sea spray. Soil dust and most industrial dusts come within this category, as do

Table 1.1 Terminal settling velocities of rigid spheres of unit density in air at 1000 mb and 200°C.

Diameter (μm)	0.1	0.4	1	4	10	40	100
Velocity (mm sec^{-1})	9×10^{-4}	7×10^{-3}	3×10^{-2}	0.5	3	50	250

pollens, mould spores and some bacterial cells. There is some consensus that, because of their high settling velocity and the unlikelihood of their entering the lung, it is not useful to consider particles of diameter greater than 100 µm.

The **fine particle mode** is a term used to indicate the combined nucleation and accumulation modes. It consists of particles less than about 1 µm in diameter.

When ambient particles are monitored by number, at most locations the overwhelming number of particles are in the fine mode. However, because volume and mass depend on (particle diameter)^3, and assuming that density is constant across the size range, it follows that 1 particle of diameter 10 µm has the same mass as 10^6 particles of diameter 0.1 µm. This results in the mass/size distribution of particulate matter being very different from the number/size distribution. At most locations, the particle mass is approximately evenly distributed between the fine and coarse modes.

The United States Environmental Protection Agency (USEPA) has made recommendations as to the appropriate size fraction to collect airborne particulate material. The agency's first national ambient air quality standards for particulate material, published in 1971, were in terms of "total suspended particulate matter" (TSP), as measured by the U.S. High-Volume Sampler. It was subsequently realised that the sampler has unsatisfactory particle-size/collection-efficiency characteristics – in particular, the particles are not closely related to the size fractions entering or depositing in the various regions of the lung. The USEPA has therefore promulgated revised standards in terms of particulate material with a diameter less than a nominal 10 µm – PM_{10}. In fact, the USEPA has defined PM_{10} to correspond with ISO's thoracic fraction. The ISO definition (ISO, 1995) of the thoracic fraction satisfies the USEPA's PM_{10} criteria, since an ideal thoracic sampler would collect 50% of particles having an aerodynamic diameter of 10 µm and would collect other sizes with efficiencies within the specifications detailed in the USEPA's 1987 recommendations. It follows that PM_{10} (or ISO's thoracic fraction) is probably the single most useful size fraction to collect when monitoring the levels of particulate material in ambient air. Both ISO and the USEPA have fully set out the reasoning behind their choices of particular size fractions. Furthermore, the USEPA's standards for particulate material in ambient air appear to be the most restrictive ones currently in use, and these are defined in terms of PM_{10}. It should be borne in mind, however, that because PM_{10} is a measure of the material entering the whole lung, it may not be the ideal measure for effects on materials.

1.6.3.1 The Sources and Composition of Atmospheric Particulate Material

The term "atmospheric particulate material" refers to all airborne particles, so it is by definition non-specific. It includes material from such diverse sources as, for example, vehicle emissions, the resuspension of surface dusts and soils and chemical reactions between vapours and gases in the atmosphere, which result in the formation of secondary particles. Therefore emission inventories of PM relate to primary sources of PM only (not secondary sources).

The principal types of primary particulate material are

Petrol and diesel vehicles, the latter being the source of most black smoke.
Controlled emissions from chimney stacks.
Fugitive emissions. These are diverse and mostly uncontrolled and include

The resuspension of soil by wind and mechanical disturbance.
The resuspension of surface dust from roads and urban surfaces by wind,
vehicle movements and other local air disturbance.
Emissions from activities such as quarrying, road and building construction, and the loading and unloading of dusty materials.

Secondary particles are those arising when two gases or vapours react to
form a substance that condenses onto a nucleation particle. The main sources of
secondary particles are the atmospheric oxidation of sulphur dioxide to sulphuric acid and the oxidation of nitrogen dioxide to nitric acid; the sulphuric
acid is present in air as droplets, the nitric acid as a vapour. Hydrochloric acid
vapour (arising mainly from refuse incineration and coal combustion) is also
present in the atmosphere, and both this and nitric acid vapour react reversibly
with ammonia (largely arising from the decomposition of animal urine) to form
ammonium salts. Sulphuric acid reacts irreversibly in two stages to form either
ammonium hydrogen sulphate or ammonium sulphate. These ammonium salts
are formed continuously as sulphur dioxide and nitrogen dioxide are oxidised,
and ammonia becomes available for neutralisation. They are therefore part of a
large-scale pollution phenomenon affecting both urban and rural areas. Most
of the accumulation mode particles in the UK atmosphere consist of ammonium salts.

1.6.3.2 Composition

A typical breakdown of the chemical composition of particulates for an urban
area is shown in Fig. 1.7 based on background PM_{10} samples collected in
London and Birmingham, UK (Harrison, Jones and Lawrence, 2004).

In general the size distributions found for the various components of PM are
related to their origins. Thus, for secondary particles (ammonium nitrate and
sulphate) and for particles formed by combustion (carbonaceous material), the
greater part of the mass is associated with particles below 2.5 μm in diameter. By
contrast, the greater part of the mass of wind blown dusts is associated with
particles larger than 2.5 μm in diameter.

Carbonaceous particulate material contains carbon in several forms: organic
(60–90%), carbonates (<5%) and particulate elemental carbon (PEC) – also
known as black carbon or graphitic carbon (Hamilton and Mansfield, 1991).
PEC is of special interest because of its adverse effects on amenity – the
reduction of visibility and the soiling of surfaces. By applying emission factors
to fuel consumption data for the UK, Hamilton and Mansfield estimated the
annual PEC emissions for the UK for the period 1971–1986. They found that
the total PEC emissions were around 17,000 or 18,000 tonnes per year during

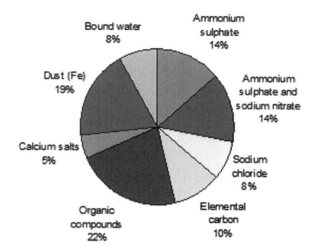

Fig. 1.7 Composition of background PM_{10} in the urban atmosphere (London and Birmingham)

most of the period, but rose during the last few years to about 22,000 tonnes. They also found that the contribution from diesel fuel consumption increased fairly steadily throughout the period – from about 80% initially to about 90% in 1986.

1.6.4 Ground-Level Ozone

Ground-level ozone (O_3) is a secondary pollutant; it is generated from the reactions of primary pollutants in the atmosphere. It is important to note that ozone at ground level is classified as a pollutant, in stark difference to strato-spheric ozone and the ozone layer occurring at around 11–15 km above ground level. Stratospheric ozone is naturally occurring and critically beneficial, as it protects us from harmful levels of UV radiation from the sun. Tropospheric or ground-level ozone is formed mainly from the breakdown of NO_x into NO_2 in the presence of UV (sunlight).

NO_2 + UV → NO + O (NO_2 plus sunlight breaks the molecules to give rise to O free radicals)

O + O_2 + M → O_3 + M (where M is any molecule, such as N_2 or O_2, ozone is produced from the joining of the free radical and other oxygen molecules)

NO + O_3 → NO_2 + O_2 (NO_2 production, the ozone is quenched to form NO_2)

This cycle of reactions generates a contribution to the ground-level ozone concentration, with photolysis giving O_3 as a by-product. Once formed, it persists and can be transported long distances. In southern England, for example ozone episodes can occur when air masses are transported from continental Europe in summertime. It is also known as a rural pollutant, because concentrations are often lower in urban areas, as direct NO emissions "scavenge" the available O_3 to form NO_2. In rural areas, the lack of NO means that the ozone molecules are not destroyed. As the reaction needs sunlight, O_3 shows a diurnal pattern with highest concentrations occurring at midday or mid afternoon. A seasonal pattern also exists with spring and summer having higher concentrations in the presence of stronger UV light.

Ground-level ozone has several effects; it causes ground-level haze, photochemical smog, vegetation damage and degradation of materials, such as rubber, dyes and books. It can increase susceptibility to respiratory infections; it is a strong pulmonary irritant causing coughing, eye, nose and throat irritation and headaches (especially when exercising).

Volatile Organic Compounds (VOCs) are precursor pollutants that lead to the formation of ozone. VOCs are an umbrella term for a class of organic compounds, which evaporate easily and contribute mainly to the production of secondary pollutants such as ozone. VOCs include hydrocarbons (alkanes, alkenes, aromatics), oxygenates (alcohols, aldehydes, ketones, acids, ethers) and halogen-containing species. They are often classed as methane (CH_4) or non-methane VOCs (NMVOC). They have a large number of varied sources: vegetation, manufacturing and industrial processes, evaporation of solvents, biogenic processes and combustion. Emission levels of NMVOC in the UK are estimated at 2.5 Mt for 1990 and 1.1 Mt for 2003, showing a decrease of 55% over these years. The observed decrease arises primarily from the road transport and industrial sectors (Fig. 1.8).

Peroxyacetyl nitrate (PAN) is another photochemical oxidant, a member of the NOy family, formed from chemical reactions involving HCs and NO_x, with a chemical formula of $C_2H_3O_5N$. It has a similar pattern to O_3, with rural concentrations of about 0.1–1.8 ppb and contributes to photochemical smog and poor visibility.

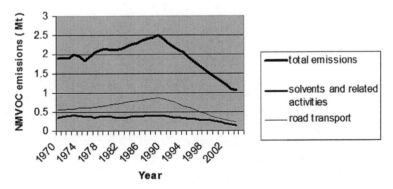

Fig. 1.8 NMVOC emissions in the UK from 1970 to 2004. (NAEI, 2008)

There are several regions in the world that suffer from damaging levels of ozone, for example in continental Europe, Los Angeles (USA) and Australia. In the latter, natural biogenic emissions of VOC from bush vegetation, e.g. volatiles (oils or the scent) from Eucalyptus trees, cause a photochemical haze and promote ozone production. The Blue Mountains in New South Wales are named for their blue haze colour produced from the eucalypts and give rise to elevated O_3 concentrations in the Sydney Metropolitan Area.

Ozone concentrations can be reduced in theory by controlling emissions of precursor pollutants (VOCs and NO_x). However, in practice, the balance between concentrations of precursor gases can increase ozone concentrations by promoting its formation.

1.7 Trends and Scenarios

Pollution levels at a specific location will change with time because of temporal variations in the emissions or the prevailing meteorology. In particular, developments in pollution control technology and air quality management have led to significant reductions in emissions levels. The main methods for emission reductions have been

Control of industrial emission
Fuel modification
Combustion modification
Flue gas cleaning

Control of traffic emissions
Fuel modification
Exhaust gas cleaning (catalytic converters, particle traps)
Removing gross polluters
Traffic management

Regulatory approaches (see also Chapter 9)
Legislation
Implementation and enforcement
Sustainability
Carbon tax

Changes in fuel use, impacts of legislation and energy-saving efficiency measures have driven the reduction in emission rates of the main air pollutants in the developed world over the past few decades. However, the effects of developments in vehicle-related technology and fuel efficiency can be offset with the increase in the number of vehicles on the road as car ownership continues to increase (Fig. 1.9) leading to a mixed picture of variation in U.S. pollutant emission levels in the past 30 years.

Industrialisation and increased development of the world economy in countries that previously had less power-hungry economies, e.g. in China and India,

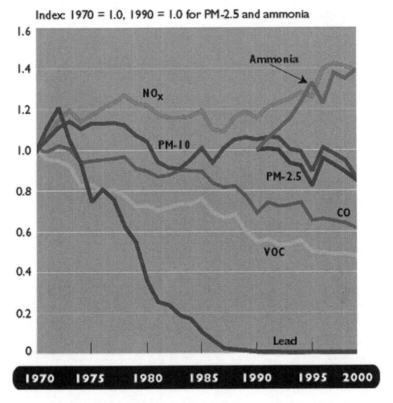

Fig. 1.9 US transportation emissions 1970–1999 related to a 1970 base level. (USEPA, 2008)

has meant that world-wide emissions have increased in recent years. On a global scale this increase will offset the reduction made in developed countries. However, legislation, such as the EU Directive on National Emissions Ceilings, will continue to drive emission rates down. The UK Government's DEFRA expects that the UK will achieve the limits on air pollution emissions required by the National Ceilings Directive in 2010 for ammonia, VOCs, SO_2 and NO_2. Table 1.2 below shows the % reduction for the main pollutants occurring during the period 1990–2003. The importance of road transport as a contributor to these pollutants, combined with coal combustion, continues to be responsible for a large proportion of sources of these pollutants today (Table 1.3).

Future emission predictions show some rises in emissions from certain sources. The European Pollutant Emission Register (EPER, 2007) highlights rises in industrial emissions that occurred from 2001 to 2004. Although industrial sulphur dioxide emissions fell by 11%, industrial growth in eastern European countries offset this reduction. The Czech Republic was the largest source of PM_{10} in 2004 of over 60 tonnes per annum, contributing to 21% of the total 2004 European emissions. In contrast the UK contributed to 6% that year.

Table 1.2 Changes in UK emissions for various pollutants from 1990 to 2003, shown as a percentage of 1990 emissions

Pollutant	PM_{10}	$PM_{0.1}$	PM_1	$PM_{2.5}$	Black Smoke	NO_x	SO_2	NMVOC
% reduction 1990–2003	51	35	42	39	66	44	74	55

Table 1.3 The contribution of road transport and coal combustion to total UK emissions in 2003

Pollutant	Coal Combustion (%)	Road Transport (%)	Total Contribution (%)
SO_2	68	0	68
NO_x	19	35	54
PM_{10}	10	25	35
Carbon	21	20	41

(NAEI, 2008).

Future trends in emissions are likely to be determined by market demand and government administration policies associated with these emission sources, e.g. transport policies, renewable energy and fuel supplies.

Acknowledgments The authors are grateful to DEFRA, EEA and USEPA for giving their permission to use information available in their reports and on their web sites.

References

Anderson HR 1999, Health effects of air pollution episodes, in: Holgate ST, Samet JM, Koren HS and Maynard RL (eds) *Air Pollution and health*, Academic Press, London; ISBN 0-12-352335-4.

AQEG 2006, Trends in Primary Nitrogen Dioxide in the UK. Draft report for comment, August 2006. available from www.airquality.co.uk

Bailey DLR and Clayton P 1982, The measurement of suspended particle and total carbon concentrations in the atmosphere using standard smoke shade methods, *Atmos. Env.* 16, 2683–2690.

Brimblecombe P 2003, *The effects of air pollution on the built environment*. Imperial College Press, London.

BSI 1969, BS1747 Methods for the measurement of air pollution, Part 2, Determination of the concentration of suspended matter. BSI London.

DEFRA 2004, Nitrogen dioxide in the United Kingdom – report by the Air Quality Expert Group (AQEG) *PB9025A*.

Friedlander SK 2000, Smoke, dust and haze: fundamentals of aerosol dynamics (second edition), Oxford University Press ISBN-13:978-0-19-512999-1.

Hamilton RS and Mansfield TA 1991, Airborne particulate elemental carbon: its sources, transport and contribution to dark smoke and soiling, *Atmos. Env.* 25, 715–723.

Harrison RM, Jones AM and Lawrence RG 2004, Major component composition of PM_{10} and $PM_{2.5}$ from roadside and urban background sites, *Atmos. Env.* 38, 4531–4538.

Hinds WC 1999, Aerosol Technology: properties, behaviour and measurements of airborne particles, 2nd edition. Wiley ISBN 978-0-471-19410-1.

ISAC 2004, Noah's Ark, Global Climate Change Impact on Built Heritage and Cultural Landscapes. http://noahsark.isac.cnr.it/

ISO 1995, Air quality – particle size fraction definitions for health related sampling, ISO 7708, International Organisation for Standardization, Geneva.

NAEI, 2008, United Kingdom National Atmospheric Emissions Inventory. http://www.naei.org.uk/

Saiz-Jimenez C (editor) 2004, *Air pollution and cultural heritage*. A.A. Balkema Publishers, Taylor & Francis Group plc, London.

USEPA 2008, US-EPA National Emissions Inventory. http://www.epa.gov/ttn/chief/net/neiwhatis.html

Sources of Additional Information

APEG (Airborne Particulate Expert Group) Particulate matter in the United Kingdom. 2005. available from http://www.defra.gov.uk

Colls J, (2002) Air Pollution (Clay's Library of Health & the Environment), 2nd edition, Spon Press, ISBN 0415255643.

EMEP programme (Co-operative Programme for Monitoring and Evaluation of the Long-range Transmission of Air pollutants in Europe) http://www.emep.int/index.html

European Pollutant Emission Register (EPER) http://www.eper.ec.europa.eu/eper

Friedrich R and Reis S (editors), 2004, Emissions of Air Pollutants: Measurements, Calculations and Uncertainties, Springer ISBN 3-540-00840-3.

Godish T (2003) *Air Quality 4th edition*, CRC Press, ISBN 9781566705868

Hewitt CN and Jackson A (editors), 2003, *Handbook of Atmospheric Science*, Wiley, ISBN 9780632052868

LAEI (2003) – London Atmospheric Emissions Inventory: http://www.london.gov.uk/mayor/environment/air_quality/research/emissions-inventory.jsp

National Academy of Sciences (2004) *Air Quality Management in the United States* , National Academy Press. ISBN 0309089328.

UK National Atmospheric Emissions Inventory. http://www.naei.org.uk

UK National Air Quality Information Archive. http://www.airquality.co.uk

Meteorological, emissions, pollution, etc. in Spain

National Level: Spanish Environment Ministry http://www.mma.es

Meteorological National Institute http://www.inm.es

City Level (Madrid): http://www.mambiente.munimadrid.es

Chapter 2
Monitoring, Modelling and Mapping

Ron Hamilton, Helen Crabbe, Stephan Fitz, and Terje Grøntoft

2.1 Overview

Chapter one discussed the role of air pollution in damaging our cultural heritage and showed that, in general, emissions have reduced but also that the pollution is changing in its nature with the evolution of a new pollutant environment. The dominance of SO_2 pollution has fallen and traffic derived pollutants have increased, creating a new multi-pollutant situation.

It is necessary to be able to estimate the level of air pollution to which the surface of a monument or other piece of heritage is exposed if one wishes to make an estimate of the potential damage that might be caused by air pollution. This chapter looks at ways that this might be done.

There are a number of useful sources of information, and methods of utilising them, that are presented here. The advent of automatic monitoring equipment together with advanced telemetry to permit the resulting data to be collected and collated centrally has enabled many countries and local authorities to install automated air quality networks. In many cases the data is available on the internet. This data, of course, represents a snapshot of the pollution situation at a given moment in time. It can be represented as a map by use of statistical interpolation.

As we have seen, emission inventories have also been developed for a number of pollutants and, combined with dispersion models and geographic information systems, they also provide an estimate of the likely level of pollution at any given location. In contrast to the monitored data, this data represents a calculation and is therefore much easier to manipulate to evaluate different policy options. If it was desirable to examine the potential of a traffic relocation scheme around a cathedral, for example, or the effects of introducing a new engine technology to the bus fleet of a city, then the relevant emissions factors could be reassessed and the new outcome mapped.

R. Hamilton (✉)
Centre for Decision Analysis and Risk Management, School of Health and Social
Sciences, Middlesex University, The Burroughs, London NW4 4BT, UK
e-mail: r.hamilton@mdx.ac.uk

J. Watt et al. (eds.), *The Effects of Air Pollution on Cultural Heritage*,
DOI 10.1007/978-0-387-84893-8_2, © Springer Science+Business Media, LLC 2009

Either of these approaches can therefore provide a heritage manager with a simple risk map to decide whether atmospheric pollution might likely be a problem. In the chapters which follow, we will discuss the types of damage that result and look at how the relationship between pollution and damage (so called dose-response functions) can be calculated. These latter techniques will permit us to attempt more sophisticated risk maps.

A number of case studies are included to show actual examples of modelling and mapping. A sophisticated urban model has been used to develop a picture of ambient pollution in Oslo (that will be used as a baseline for mapping soiling later in the book). National maps of ambient pollution from emission inventories and models, verified by monitoring, will be presented from the UK at two different scales – nationwide and city scale (in London). Estimates of the amount of pollution that actually deposit from the airborne concentrations will also be demonstrated for the UK.

2.2 Monitoring

An accurate assessment of the effects of air pollution on a specific unit of cultural heritage requires knowledge of the level of air pollution to which the surface materials of the heritage are exposed. Until the mid 1990s, this information could normally only be obtained by the establishment of a monitoring programme to record the levels of pollution; a time-consuming and resource-intensive (and consequently costly) exercise. However, an improved awareness of the adverse effects of air pollution particularly on health, combined with developments in information technology, has resulted in a large increase in the level of automatic monitoring. For example, Fig. 2.1 shows the increase in the

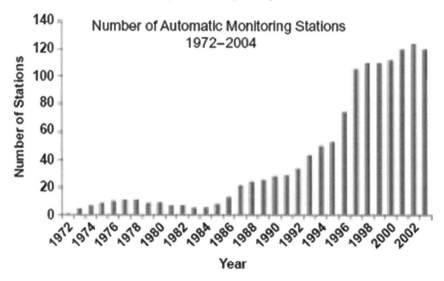

Fig. 2.1 The increase in the number of automatic monitoring stations with time (UK)

number of monitoring stations with time in the UK (UK National Air Quality Information Archive, 2008)

Currently, in most regions, the combination of a network of automatic monitors and a dispersion model capable of predicting air pollution levels at a specific location has meant that a local monitoring programme is no longer required. There are, however, some exceptions to this general rule. The location and/or nature of the unit of cultural heritage may be such that there could be significant differences between the pollution levels at that unit and those in the surrounding area. Also, since it is the exposure of the monument surface which is of importance, it is *deposition rate* rather than *ambient concentration* which is the important parameter. The deposition rate can be estimated from the ambient concentrations via a deposition velocity as discussed later in this chapter.

2.2.1 Monitoring Networks for Ambient Air

A typical monitoring network is shown schematically in Fig. 2.2.

The principles and operational characteristics of the instrumentation used in a monitoring station are extensively covered in the literature (Patnaik, 1997; Reeve, 2002; Robinson, 2003). Table 2.1 gives a summary of the main types of instrument used in an automatic air quality monitoring station.

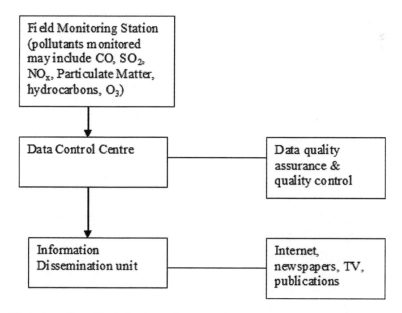

Fig. 2.2 Automatic ambient air monitoring network

Table 2.1 Measurement techniques for automatic analysers

Pollutant	Measurement technique	Response time	Minimum detectable concentration
Carbon monoxide	Non-dispersive infrared (NDIR)	5 s	0.5 ppm
Nitrogen oxides	Chemiluminescent reaction with ozone	1 s	0.5 ppb
Ozone	Chemiluminescent reaction with ethane	3 s	1 ppb
Sulphur dioxide	Hydrogen flame photometry	25 s	0.5 ppb
	Pulsed fluorescence	2 min	0.5 ppb
PM_{10}	Tapered element oscillating microbalance (TEOM)	15 min	$1.5\ \mu g\ m^{-3}$

2.2.1.1 Automatic Monitoring Networks in Europe

It is a requirement of all European Union (EU) countries to monitor and report air quality within their country. EU Directives (Chapter 9) specify the pollutants which must be monitored and provide guidance on the appropriate methods for meeting this requirement. It is the responsibility of each country to establish its network to meet EU requirements. There may also be additional national requirements as determined by the country. Results for all countries are reported to the European Environmental Agency which then integrates and disseminates results on a European scale.

The European Environment Agency (EEA) is the formal agency of the *European Union* responsible for collating the national network measurements, and for monitoring and assessing the European environment. It was established by EEC Regulation 1210/1990, as amended by EEC Regulation 933/1999; and became operational in 1994. Its headquarters are in Copenhagen, Denmark. All member states of the European Union are automatically members; and other states may become its members by means of agreements concluded between them and the European Community. Currently (2008), the EEA has 32 members:

27 EU member states
3 European Economic Area members: Iceland, Norway, Liechtenstein
1 candidate country: Turkey
Switzerland since April 1, 2006
In addition Albania, Bosnia-Herzegovina, Croatia, Macedonia, Serbia and Montenegro participate in Eionet (see below) and the EEA work programme in anticipation of becoming member countries.

Through the European Environment Information and Observation Network (Eionet, 2008), the European Environment Agency (EEA) provides data and information on the state of the environment in Europe. The State of the Environment Reporting Information System (SERIS) is an inventory of national State of the Environment reports. The database contains brief publication details for recent (1997 onwards) reports for countries included in the EEA. SERIS contains links to the reports and to the main organisation involved in the report. Reports can be identified by their geographic coverage or the year of publication. National and multi-national reports are included in the database. Eionet has five specialist Topic Centres, including the European Topic Centre on Air and Climate Change (ETC/ACC).

2.2.1.2 Case Study – Monitoring in the UK

For the purpose of monitoring and reporting air pollution, the UK has been divided into regions (or zones) and urban areas (or agglomerations), in accordance with EC Directive 96/62/EC. There are sixteen regions defined for reporting levels of air pollution. They match the boundaries of England's Government Offices for the Regions and the boundaries agreed by the Scottish Executive, National Assembly for Wales and Department of the Environment in Northern Ireland.

There are two automatic networks in the UK. These are:

The Automatic Urban and Rural Network (AURN). This presently consists of 123 sites and remains the most important single monitoring programme in the UK today. This network presently includes 87 urban, 22 rural and 14 London Network sites. The locations include background locations representing the exposure of the population for significant periods of time and 'hotspots' monitoring at urban roadsides and kerbsides and around industrial sources. Each site monitors some or all of the following:

carbon monoxide;
ozone;
sulphur dioxide;
nitrogen oxides; and
particles (as PM_{10} and sometimes $PM_{2.5}$).

The Hydrocarbon Network; this currently (since 2002) consists of 5 sites, in Cardiff, Glasgow, Harwell, London Eltham and London Marylebone Road. It monitors 32 volatile organic compounds. An earlier hydrocarbon monitoring programme operated from 1996 at more (13) sites but with recording fewer (26) hydrocarbons.

There is a substantial number of other continuous monitoring sites in the UK that are not part of the formal national network. This includes monitoring conducted by local authorities, environmental consultants and researchers. The contribution from London is described later. Most air quality data sets or summaries of monitoring campaigns are available free of charge from public bodies. The location of the automatic monitoring sites is shown in Fig. 2.3.

Because of its size and population density (7.5 million inhabitants), London has developed a network of monitoring sites to coordinate and improve air pollution monitoring in the city. The London Air Quality Network (LAQN)

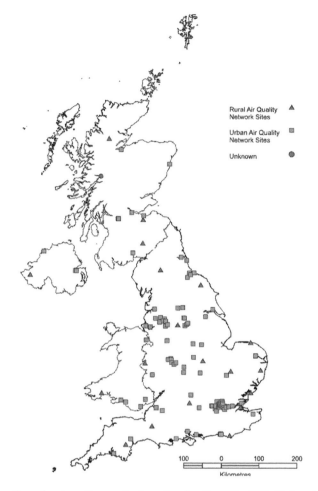

Fig. 2.3 Location of automatic monitoring sites for the UK. (UK National Air Quality Information Archive (NAQIA) 2008

was formed in 1993 to coordinate and improve air pollution monitoring in London. By 2008, 31 London Boroughs were supplying data to the LAQN. Increasingly, these data are being supplemented by measurements from local authorities surrounding London, thereby providing an overall perspective of air pollution in South East England. Fourteen of the LAQN sites are also affiliated to the national Automatic Urban and Rural networks (AURN). There are currently over 200 continuous analysers in use in the London Air Quality Network, located at over 100 monitoring sites.

In addition to the automatic networks, there are twelve non-automatic networks that measure a wide range of pollutants. The twelve are:

- Acid Deposition
- Acid Waters
- Ammonia
- UK Heavy Metals Monitoring Networks
- Nitric Acid - part of Acid Deposition network
- Nitrogen Dioxide Diffusion Tube
- Rural Nitrogen Dioxide - part of Acid Deposition network
- Rural Sulphur Dioxide and Sulphate
- Smoke and Sulphur Dioxide
- Toxic Organic Micro Pollutants
- Non-Automatic Hydrocarbon Network
- Polycyclic Aromatic Hydrocarbons

In total, there are more than 1,500 national air quality monitoring sites across the UK. Results are available as measured concentrations or as indicator values (air pollution is described on a scale of 1–10 where 1 corresponds to 'Low' pollution and 10 corresponds to 'Very High' pollution). Current, archived and forecast results are available.

2.2.1.3 Automatic Monitoring Networks in USA

The US Environmental Protection Agency (USEPA) coordinates the ambient air quality monitoring program which is carried out by state and local agencies. There are three major categories of monitoring stations, State and Local Air Monitoring Stations (SLAMS, see Fig. 2.4), National Air Monitoring Stations (NAMS, see Fig. 2.5), and Special Purpose Monitoring Stations (SPMS), that measure the criteria pollutants. Additionally, a fourth category of a monitoring

Fig. 2.4 The USA State and Local Monitoring (SLAMS) network

Fig. 2.5 The USA National Air Monitoring (NAMS) network

station, the Photochemical Assessment Monitoring Station (PAMS), which measures ozone precursors (approximately 60 volatile hydrocarbons and carbonyl) has been required by the 1990 Amendments to the Clean Air Act.

The NAMS (1,080 stations) are a subset of the SLAMS network with emphasis being given to urban and multi-source areas. In effect, they are key sites under SLAMS, with emphasis on areas of maximum concentrations and high population density.

2.2.2 Deposition Monitoring

The impact of air pollution on building follows the deposition and accumulation of the pollutants on the surface of the building and subsequent physical and chemical interactions. For this reason, deposition rate might be considered to be a more relevant parameter than ambient atmospheric concentration when assessing the consequences of the atmospheric conditions on buildings. However, deposition rate is not routinely measured while ambient atmospheric concentration is automatically measured and reported extensively as described above. For this reason, dose-response functions (Chapters 3 and 4) are normally formulated in terms of ambient atmospheric concentration and deposition rates estimated, as required, by an established methodology as described below.

Deposition is normally modelled using the following formulation:

Dry deposition flux i.e. deposition without the presence of rainfall, $D_{dry} = Ca \times v_{dd}$ where Ca is atmospheric concentration ($\mu g\ m^{-3}$) and v_{dd} is the dry deposition velocity (ms^{-1}).

Wet deposition flux i.e. deposition in the presence of precipitation, $D_{wet} = Cr \times R$ where Cr is concentration in precipitation ($\mu g\ m^{-3}$) and R is precipitation intensity (ms^{-1}). Rainfall is the main contributor to precipitation though snowfall may also be significant in some locations.

Total deposition flux $D_{tot} = D_{dry} + D_{wet}$
All fluxes are measured in $\mu g\ m^{-2}\ s^{-1}$.

The direct measurement of deposition to the surface of cultural heritage material is almost impossible because the surface cannot be invasively analysed or moved. Other than measurements based on radionuclides following the Chernobyl incident (Nicholson, 1989), deposition measurements have involved the use of surrogate surfaces. A number of studies have used materials similar in composition to the materials present in cultural heritage with these surrogate materials mounted in a rack (Fig. 2.6) and exposed to the atmosphere at a location in close proximity to the cultural heritage building under investigation (Ferm et al., 2006). Laboratory studies developed by a team of scientists from the University of Vienna (Aksu et al., 1996) involved placing cubes in a wind tunnel to simulate ambient conditions (see also Chapter 4). This arrangement allowed them to measure dry deposition of particles to building surfaces and the soiling which followed.

There is a substantial literature on the deposition of air pollutants to other materials; however, deposition rate is a function of the physical nature of the surface receiving the depositing material, the local micrometeorology and, in the case of particulate matter, the physical and chemical properties of the particles. For this reason, recorded deposition rates vary substantially. For example, in a review of published values, McMahon and Denison (1979) showed that reported v_{dd} for SO_2 to ecosystems varied between 0.3 and 1.6 cm s^{-1}. Reviews of deposition velocity measurements and the relative importance of wet and dry deposition have been given by Matt et al. (1987), Nicholson (1988) and Lipfert (1989).

Simultaneous measurements of climatic parameters – temperature, relative humidity, amount and acidity of precipitation – and pollution parameters – SO_2, NO_2, HNO_3, O_3 and particles as PM_{10} gives the data for assessment of the

Fig. 2.6 Sampling racks in Prague (a) and Tehran (b) for investigating deposition and effects on materials

Fig. 2.7 Passive samplers for gaseous pollutants (*right*) and particulate matter (*left*)

general level of damage to materials in a particular area. Measurement of pollution parameters can be performed using active sampling with special, usually quite expensive, devices, or passive sampling with small samplers (Fig. 2.7). These are ideal for sampling around objects of cultural heritage because the sampling is silent, does not need electricity, can be used for long-term integrative sampling and can be performed inconspicuously and with discretion. Other advantages are that technical personnel are not needed for exposing the samplers and that they do not need field calibration. A surrogate surface exposed in the vicinity of the object of interest receives a similar net particle or gas deposition as, for example, an object of cultural heritage. Use of passive samplers in this way is also desirable because it replicates the complexity of particle or gas deposition to actual objects, which depends on many parameters such as particle concentration, its size distribution, wind speed, turbulence and surface characteristics of the object in a complicated way. This type of standardised exposure on test sites does not, however, take into account possible effects of microclimate, which may occur in different positions on buildings and monuments.

Most networks for the routine monitoring of deposition resulted from concerns over the effects of acid rain on ecosystems. In the UK, wet deposition is monitored at 38 sites across the country through the collection and analysis of rainfall samples using bulk collectors. Measurements from five of these sites form the UK contribution to the European Monitoring and Evaluation Programme (EMEP) network. The rainfall sample is used to estimate rain ion concentrations at each site and also, for sulphur and base cations, to identify the proportion of the concentration coming from sea salt as opposed to other, generally anthropogenic, pollutant emissions. Along with rainfall maps for the UK, these data are used to estimate the wet deposition of non-sea salt sulphate

(SO_4^{2-}), nitrate (NO_3^-), ammonium (NH_4^+) and non-sea salt base cations (Ca^{2+}, Mg^{2+}) across the country. Cloud droplet deposition, the removal by vegetation of fine mist droplets (for example when a high elevation forest plantation is covered by hill fog), is generally a small component of acid deposition but is also estimated from the rain ion concentrations.

Dry deposition of sulphur and nitrogen to vegetation and other surfaces is a more complex process, with the possibility of uptake of gases by several pathways such as through stomatal pores, onto plant surfaces and by absorption in leaf water films. Gas concentrations $(SO_2, NO_2, NH_3, HNO_3)$ are monitored at varying numbers of rural sites across the UK and these data are input to models of surface-atmosphere exchange to derive estimates of the dry deposition of sulphur and nitrogen. A similar procedure is used to map the deposition of base cation particles.

As described later in this chapter, deposition across the UK is currently mapped on a 5 km × 5 km square grid basis by interpolating maps from the measured pollutant concentrations and using models that also incorporate spatial information on meteorology and land use. These maps do not provide deposition estimates to point locations or small areas, which will often be substantially different from the average value in the 5 km square, as the major causes of this local variation are not represented within the national mapping programme.

2.3 Modelling

Cultural heritage buildings and objects exposed outdoors in urban environments are exposed to elevated concentrations of most pollutants as compared to those exposed in the rural background. Structures located at hot spots (e.g. traffic sources) in cities are usually even more exposed than structures located in the urban background. An exception from this general situation is ozone (O_3), which in Europe usually has lower concentrations in urban than in rural locations and at urban hotspots as compared to the urban background, due to its reaction with NO emissions from combustion processes, and especially traffic sources. As a consequence of the increased pollution, increased atmospheric degradation of cultural heritage is observed in urban areas. The large focus on urban air quality for the safeguarding of people's health should therefore be complemented with assessments of the effect of reduction of air pollutants on the preservation state of exposed cultural heritage structures and objects.

Urban air quality can be predicted by using different atmospheric chemistry transport dispersion models which are able to reflect changes in urban emissions (Chock and Carmichael, 2002; Vawda, 2003). A host of different models have been developed and used to study conditions in different cities. The models have been developed mainly as a response to health authorities' need to know if air quality standards are met in urban regions with elevated pollution levels. The dispersion models make it possible to evaluate the distribution of air pollution in urban areas, often with a high resolution. They also make evaluation of the exposure of the population to the pollutants possible, and also allow evaluation of the effect of abatement strategies aimed to meet air quality standards to reduce emissions. The models also have general applicability for effects studies. They are well suited to assessing the elevated doses of pollutants that urban cultural heritage buildings and objects are often exposed to, and to compare doses with tolerable levels of exposure. The effect of abatement strategies can be investigated, as they have been for health purposes. Output from dispersion – abatement models for localities of interest can be used as input to outdoor-to-indoor models to evaluate expected indoor exposure (see Chapter 8). Health standards and indicators are partly different from those related to effects on cultural heritage materials, structures and objects. The number of daily exceedances of critical values is usually part of health standards. For cultural heritage long-term integrated doses measured as, for example, yearly mean concentration values are used. The output from dispersion modelling can be formulated to whichever of these purposes is of interest.

The European Topic Centre on Air and Climate Change (ETC/ACC) assists the European Environment Agency (EEA) in its support of European Union (EU) policy in the field of air pollution and climate change. It is a consortium of 14 European institutions, established in 2001 by the EEA. The ETC/ACC reports on the progress of EU environmental policy on air pollutant emissions, air quality, and climate change issues. The ETC/ACC also maintains an online copy of the Model Documentation System (MDS, 2008) which is a catalogue of most of the air quality and air pollution dispersion models developed and/or used in Europe. The MDS catalogue listings include the name, description and support contacts for each model as well as other pertinent technical details. The MDS, developed at the Aristotle University in Thessaloniki, Greece, contains information on 119 models and aims to provide guidance to any user in the selection of the most appropriate model for his application. Inclusion of a model in the system does not imply endorsement for using the particular model but it provides users with the option to assess the most appropriate model by using the specifications submitted by the original modellers.

On the European level the EMEP (EMEP 2008) model is the dominant model, though at national and local levels a range of different models are used, which include different parameters with different geographical resolution, based both on measurements and on emission inventories.

An illustration of the functionality and typical use of dispersion and abatement modelling, emphasizing the methodology, is given below for model calculations with the AirQuis system for the city of Oslo, Norway. Similar exercises to model annual mean values of pollutants and subsequently assess different degradation effects on cultural heritage can be performed wherever the modelling resources are available.

2.4 Case Study – Model Calculations to Estimate Urban Levels of Particulate Matter in Oslo. Methodology and Model Evaluation

During the winter and spring seasons, Norwegian cities are susceptible to poor air quality events that can lead to concentrations in exceedance of limit values. Such events typically take place during periods with strong temperature inversions, weak winds and little vertical mixing. For particulate matter, these episodes are enhanced during cold and dry conditions during the winter and spring when emission of particulates from domestic wood burning and from traffic induced resuspension are at their highest. This can lead to concentrations in exceedance of health limit values, as defined in the Council Directive 1999/30/ EC and contribute to increased soiling of buildings. Modelling studies of PM_{10} and $PM_{2.5}$ for 2003 and model simulations of ambient PM_{10} and $PM_{2.5}$ concentrations, as well as population exposure, in 2010 and 2015 were performed for Oslo as part of a larger study intended to assess the impact of recommendations made in connection with Revision 1 of the EC Daughter Directives.

The modelling system applied to simulate concentrations of particulates in this study is the AirQuis modelling system, developed at NILU (AirQuis, 2004; MDS, 2008). AirQuis is a PC-based integrated management system that includes a user interface, an extensive database solution, a comprehensive emission module, a suite of models for use in simulations, exposure models and a GIS-based system for presentation and analysis.

The models used in the calculations are the MATHEW diagnostic wind field model (Sherman, 1978; Foster et al., 1995) and the EPISODE dispersion model (Slørdal et al., 2003). This dispersion model contains a standard Eulerian type model (i.e. one that uses a fixed three-dimensional Cartesian grid to track the movement of a large number of pollution plume parcels as they move from their initial location) for calculating concentrations from area emissions as well as the line source model HIWAY-2 (Petersen, 1980) which is used to calculate traffic related contributions at receptor points close to roads.

The grid applied in the Oslo region is a 22 × 18 km grid, grid size 1 km, with 10 vertical levels up to 2400 m. The meteorological field is calculated with MATHEW using input from a meteorological mast. Calculations are carried out for a period of 1 year, 2003, and results at receptor points corresponding to the positions of monitoring stations in Oslo are recorded for comparison with the observations recorded. Average concentrations and exceedance fields are also used for mapping and exposure calculations. Background concentrations for the model are taken from a regional background station.

Emission from traffic is introduced into the model in two separate ways, namely as area or as line source emissions, dependent on the average daily traffic intensity. The Eulerian model is applied to calculate the concentration levels in the model grid system and at individual receptor points. At receptor points close to main roads, where most monitoring stations are placed, the

HIWAY-2 line source model is applied to estimate the contribution from the nearest roads. The contribution of road related sources greater then 500 m from the receptor points is calculated solely through the Eulerian model.

The emission model for traffic-induced resuspension of particulates is dependent upon traffic speed, heavy-duty vehicle fraction, percentage of studded tyres and road surface condition. Due to the difficulty, but importance, in determining the state of the road surface (dry/wet) from available meteorological data, the surface condition is defined based on meteorological and monitoring station data (e.g. temperature, dew point temperature, precipitation).

Emissions of PM from wood burning are introduced into the model as area sources into the three lowest levels of the Eulerian model (71 m). These emissions are based on factors derived from wood consumption, fireplace type etc (Finstad et al., 2004).

Measurements of PM_{10} and $PM_{2.5}$ are carried out at 7 stations in Oslo using TEOM instruments. Hourly averages are available from two of these stations for the entire 2003 period. These stations, Kirkeveien and Løren, are analysed in this study and are representative of traffic/wood burning and traffic stations respectively. In 2003 the number of days when the daily average concentration of PM_{10} exceeded 50 $\mu g\, m^{-3}$ was 37 and 60 respectively at these two stations.

Results of the model calculations have been compared to observed levels of PM_{10} at the two stations Kirkeveien and Løren for the year 2003. The calculated daily average concentrations of PM_{10} are shown, in order of descending concentration, in Figs. 2.8 and 2.9 for both, observations and model calculations. For the station at Løren there is good agreement between model and measurements though not all peak values are captured and the lower concentration levels, corresponding mostly to summer time periods, are also underestimated. However

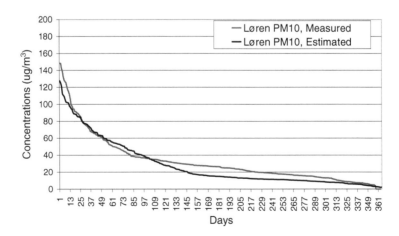

Fig. 2.8 Unpaired daily average concentrations of PM_{10} ordered in descending concentration for the Løren station

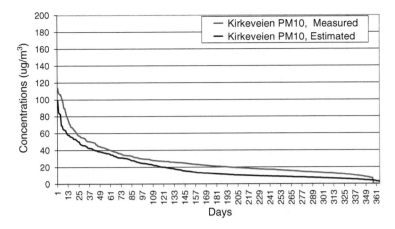

Fig. 2.9 Unpaired daily average concentrations of PM_{10} ordered in descending concentration for the Kirkeveien station

the number of exceedances is well modelled. At Kirkeveien the daily concentrations are slightly underestimated.

The ability of the model to correctly simulate particulate concentrations is dependent on 3 main aspects; the models description of meteorology, of emissions and of transport processes.

For traffic related stations the correlation between observed and calculated concentrations is to a large extent dependent on the timing of traffic. Meteorological information, used in the model, also influences correlation. The station Løren, which is situated closest to the meteorological mast used as meteorological input, shows the best correlation of all the stations in Oslo for all pollutants. It is clear that much can be gained by improvements in the resuspension model. The emission intensity of particles is currently empirically adjusted in this model, as the data required for establishing road wetness is not available and the complexity of road surface conditions makes objective determinations difficult. The model used to determine near road concentrations is an open line source model. In reality most stations are only partially 'open' and many are significantly influenced by nearby buildings and other obstacles. The largest deviations between calculated and observed values often occur when wind directions are not properly defined for the line source model.

The model evaluation showed that the AirQuis system was able to reproduce observed PM_{10} concentrations rather well. The system was consequently considered well suited for studies of air quality exceedances of the proposed limit values of the EC Daughter. On request by the Norwegian Pollution Control Authority, NILU has examined the consequences for Norwegian cities of the various recommendations made in connection with the first revision of these directives.

2.5 Mapping

The development of sophisticated computer systems, hardware and software, together with the reduced cost of these systems has encouraged the development of information technology-based systems for environmental monitoring, data collection and novel methods of presentation. The presentation of the information in the form of maps has many benefits. It is readily understandable by populations and has high visual impact. It can be adapted to a wide range of spatial scales (global/national/regional/local) and frequently offers the viewer the opportunity to 'zoom' in or out of the map (ERG, 2008).

Geographic information systems (GIS) are being used increasingly for the presentation of environmental data (Goodchild et al., 1996). Many countries now routinely provide pollution maps in publications and on the internet, with these maps showing current levels of pollution and, in some cases, predicting future levels. The EU-funded SAVIAH project developed a regression-based air pollution model in a GIS application (Briggs et al., 1997). The pollution model used data collected in, and GIS environments designed around, the European cities of Amsterdam, Huddersfield, and Prague. The resulting model was able to provide accurate pollution predictions. In addition to providing information on pollution, maps can be used to show the effects of pollution on health and ecosystems. This is achieved by combining the pollution information with risk information (e.g. dose-response functions) within the GIS. The risk information needs to include population distribution for health risk assessment, and stock at risk distribution for materials risk assessment. These approaches add a further dimension to risk assessment (Vine et al., 1997). GIS systems are commonly used for assessing exposure to air pollution (Cyrys et al., 2005). Overlaying maps of exposure and populations may define populations at risk. However, linking exposure to disease is highly dependent on the accuracy of exposure assessment as well as the time elapsed between initial exposure and disease (the latency time). The role of GIS in exposure assessment is discussed in detail by Nuckols et al. (2004).

This chapter provides some examples of air pollution mapping. Applications to corrosion, soiling and stock at risk are presented in later chapters.

2.6 Case Study – Current Levels of Ambient Pollution at a National (UK) Scale

The information recorded by the monitoring sites, when combined with a dispersion model, allows for the production of maps showing the spatial distribution of pollution across a region. Figure 2.10 shows a map of SO_2 pollution for the UK at background sites and clearly shows the impact of power station and industrial emissions in NW England, the Thames Estuary and River Clyde/

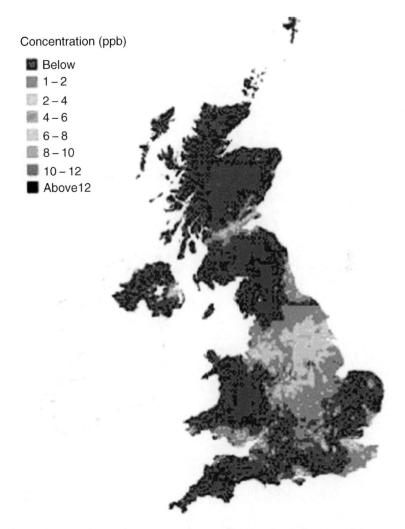

Concentration (ppb)

Below
1 – 2
2 – 4
4 – 6
6 – 8
8 – 10
10 – 12
Above12

Fig. 2.10 Background SO_2 in the UK (2002). (UK National Air Quality Information Archive NAQIA 2008)

Firth of Forth area of Scotland, as well as domestic emissions focussed around Belfast in Northern Ireland.

Although some NO_2 is emitted directly from vehicles or other sources, most is formed by rapid chemical reaction in the atmosphere. Concentrations of nitrogen dioxide therefore tend to be highest in urban areas where traffic densities are high, such as in London. The same is true for PM_{10} concentrations. Although the data mapped in Figs. 2.11 and 2.12 are background rather than roadside pollution levels, they clearly follow closely the country's major motorways and road network infrastructure.

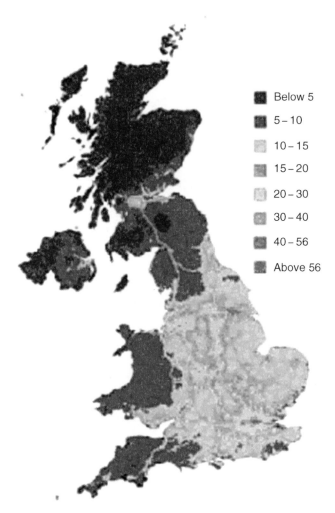

Fig. 2.11 Background NO$_2$ in the UK 2002. Concentrations are annual averages in μg m^{-3}

The ubiquitous nature of PM$_{10}$ concentrations across the country is a reflection of the range of emitting sources and the relatively strong impact of secondary sources.

As described in Chapter 1, ground-level ozone (Fig. 2.13) is formed by a series of chemical reactions involving oxygen, oxides of nitrogen and hydrocarbons in the presence of ultraviolet radiation. Because of the time required for ozone formation, high concentrations can often be formed considerable distances downwind of the original pollution sources. For this reason concentrations in busy urban areas are often lower than in the surrounding countryside.

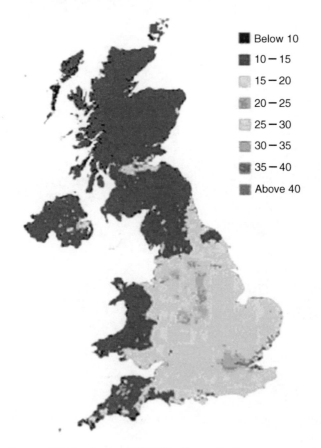

| Below 10 |
| 10 – 15 |
| 15 – 20 |
| 20 – 25 |
| 25 – 30 |
| 30 – 35 |
| 35 – 40 |
| Above 40 |

Fig. 2.12 Background PM_{10} in the UK 2002. Concentrations are annual averages in $\mu g\ m^{-3}$

2.7 Case Study – Current Levels of Ambient Pollution at an Urban (London) Scale

London provides a good case-study for the application of the same methodology (emissions inventory combined with a dispersion model and data from automatic monitoring stations to provide a regional map) applied to an urban area as opposed to the background sites described above. The Greater London Author-ity (GLA) has prepared an emissions inventory which covers the area inside the London outer orbital road, the M25. The London Atmospheric Emissions Inventory (LAEI) includes emissions from air pollutant sources of the UK Air Quality Strategy (AQS) pollutants. The city was divided into 1 km × 1 km grids

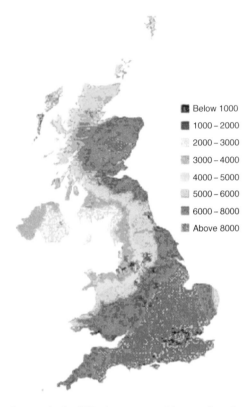

Fig. 2.13 Background ozone in the UK, shown as total hours when the ozone concentration was above 80 μg m^{-3}

and the emissions from all traffic, industrial and domestic activities within that grid were calculated. Table 2.2 shows that for London, traffic is the dominant source of PM$_{10}$.

The emissions from each grid can be combined with a pollution dispersion model to yield a contour map of pollution in the city. Figure 2.14 shows the PM$_{10}$ contour map for London.

Table 2.2 PM$_{10}$ emission totals for London (Tonnes yr^{-1})

Source type	1999	2004
Road	3,073 (76%)	2,029 (68%)
Non-road	968 (24%)	967 (32%)
Total	4,041	2,996

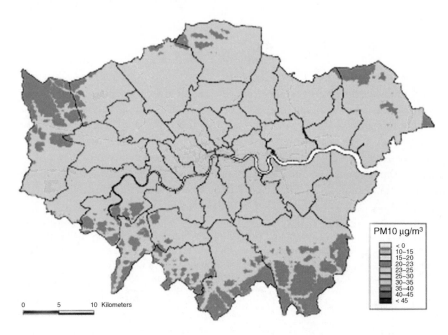

Fig. 2.14 Annual mean PM_{10} concentrations across, Greater London, 2004, modelled using ADMS Urban

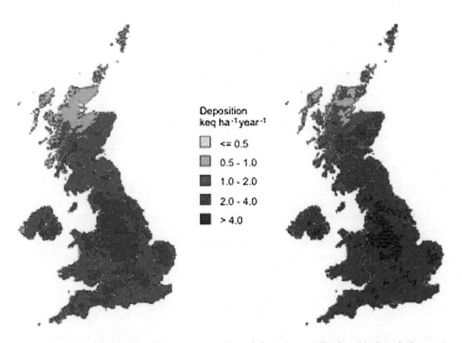

Fig. 2.15 Levels of acid deposition (non-marine sulphur plus oxidised and reduced nitrogen) for 2002–2004, UK: a) Moorland; b) Woodland. Source: Centre for Ecology and Hydrology

2.8 Case Study – Current Depositing Pollutants at a National (UK) Scale

Maps of annual mean total acid deposition (i.e. non-marine sulphur plus oxidised and reduced nitrogen) to specific ecosystems for the years 2002–2004 are shown in Fig. 2.15. The mapped values are the sum of wet, dry and cloud droplet deposition. The deposition to woodland is greater than the deposition to moorland in the same area, because the increased atmospheric turbulence associated with trees provides a route to deliver larger amounts of pollutant gases and particles from the atmosphere to the vegetation surfaces. Total acid deposition is generally greatest over England with higher deposition occurring particularly over the Pennines, Lake District, and Snowdonia.

Acknowledgments We are grateful to Thomas Gauger for his contributions to this chapter, and in particular for the mapping studies which he undertook as part of the CULT-STRAT project. We would like to acknowledge the enormous contribution of Jan Henriksen, formerly of NILU, Norway. David Dajnak and Sean Beevers (Environment Research Group, King's College London) also made valuable contributions with their studies on pollution and soiling maps for London.

References

AirQuis (2004) http://www.nilu.no/airquis/
Aksu R, Horvath H, Kaller W, Lahounik S, Pesava P and Toprak S (1996) Measurements of the deposition velocity of particulate matter to building surfaces in the atmosphere. *J. Aerosol. Sci.* 27, S675–676.
Briggs DJ, Collins S, Pryl K, Smallbone K and van der Veen A (1997) Mapping urban air pollution using GIS: a regression-based approach. *Int. J. Geogr. Inf. Sci.* 11(7), 699–718.
Chock DP and Carmichael GR (2002) *Atmospheric Modelling*, Springer, ISBN 9780387954974.
Cyrys J, Hochadel M, Gehring U, Hoek G, Diegmann V, Brunekreef B and Heinrich J (2005) GIS-based estimation of exposure to particulate matter and NO_2 in an urban area: stochastic versus dispersion modelling. *Environ. Health. Perspect.* 113(7), 987–992.
EIONET (2008) European Environment Information and Observation Network. http://www.eionet.europa.eu
EMEP (2008) Convention on Long-range Transboundary Air Pollution Steering Body to the Cooperative Programme for Monitoring and Evaluation of the Long-range Transmission of Air Pollutants in Europe. http://www.unece.org/env/lrtap/emep/welcome.html
ERG (2008) Environment Research Group, King's College London, 3-D map of air pollution in London. http://www.londonair.org.uk/london/asp/virtualmaps.asp
Ferm M, Watt J, O'Hanlon S, De Santis F and Varotsos C (2006) Deposition measurement of particulate matter in connection with corrosion studies. *Anal. Bioanal. Chem.* 384, 1320–1330.
Finstad A, Flugsrud K, Haakonsen G and Aasestad K (2004) Consumption of firewood, pattern of consumption and particulate matter. Oslo-Kongsvinger (SSB-rapport 2004/5) (available in Norwegian only).
Foster F, Walker H, Duckworth G, Taylor A and Sugiyama G (1995). User's guide to the CG-MATHEW/Adpic models, Version 3.0. Lawrence Livermore National Laboratory (Report UCRL-MA-103581 Rev. 3).

Goodchild MF, Steyaert LT, Parks BO, Johnston C, Maidment D, Crane M and Glendinning
   S, editors (1996) *GIS and Environmental Modelling: Progress and Research Issues.* Wiley,
   ISBN 978-0-470-23677-2.
Lipfert FW (1989) Dry deposition velocity as an indicator for SO_2 damage to materials.
   JAPCA 39, 446–452.
Matt DR, McMillen RT, Womack JD and Hicks BB (1987) A comparison of estimated and
   measured SO_2 deposition velocities. *Water Air Soil Pollut.* 36, 331–347.
McMahon TA and Denison PJ (1979) Empirical atmospheric deposition parameters –
   a survey. *Atmos. Env.* 13, 571–585.
Nicholson KW (1988) The dry deposition of small particles: a review of experimental
   measurements. *Atmos. Env.* 22(12), 2653–2666.
Nicholson KW (1989) The deposition, resuspension and weathering of Chernobyl derived
   material in the UK. *J. Radiol. Prot.* 9(2), 113–119.
Nuckols JR, Ward MH and Jarup L (2004). Using geographic information systems for
   exposure assessment in environmental epidemiology studies. *Environ. Health. Perspect.*
   112(9), 1007–1015.
Patnaik P (1997) *Handbook of Environmental Analysis.* Lewis, ISBN 9780873719896.
Petersen WB (1980) User's guide for Hiway-2: A highway air pollution model. Research
   Triangle Park, NC., U.S. Environmental Protection Agency (EPA-600/8-80-018).
Reeve RN (2002) *Introduction to Environmental Analysis*, Wiley ISBN 9780471492955.
Robinson R (2003) Atmospheric Monitoring Techniques. In Hewitt CN and Jackson A
   (editors), *Handbook of Atmospheric Science.* Wiley, ISBN 9780632052868.
Sherman CA (1978) A mass consistent model for wind fields over complex terrain. *J. Appl.
   Meteorol.* 17, 312–319.
Slørdal LH, Walker SE and Solberg S (2003) The urban air dispersion model EPISODE
   applied in AirQUIS$_{2003}$. Technical description. Norwegian Institute for Air Research,
   Kjeller (NILU TR 12/03).
UK National Air Quality Information Archive (2008) http://www.airquality.co.uk
Vawda Y (2003) Pollutant Dispersion Modelling. In Hewitt CN and Jackson A (editors),
   Handbook of Atmospheric Science. Wiley, ISBN 9780632052868.
Vine MF, Degran D and Hanchette C (1997) Geographic information systems: their use in
   environmental epidemiological research, *Environ. Health. Perspect.* 105(6), 598–605.

Sources of additional information

UK Defra, Environmental protection http://www.defra.gov.uk/environment/statistics/air
   qual/index.htm
US Environmental Protection Agency, Air Quality http://www.epa.gov/airtrends/
The London Air Quality Network, LAQN, http://www.londonair.org.uk
THE EUROPEAN ENVIRONMENT, STATE AND OUTLOOK 2005 European Environment
   Agency, Kongens Nytorv 6,1050 Copenhagen K Denmark. http://www.eea.eu.int
European Topic Centre on Air and Climate Change http://air-climate.eionet.europa.eu

Chapter 3
Corrosion

Johan Tidblad, Vladimir Kucera, and Susan Sherwood

3.1 Overview

Corrosion is a familiar concept – as familiar as the rusting of steel left outside or the green patina of an old copper roof. Corrosion attack is normally seen as a non-desirable effect that causes a loss of aesthetic value and mechanical strength, although many find the patina attractive. This chapter takes those simple concepts and expands them to present the actual mechanisms involved and to relate them to what is happening in the atmosphere.

The first section outlines the major corrosion processes that happen when materials are degraded. Atmospheric corrosion is complex with physical, chemical and biological processes, or a combination of them, operating. The focus of the chapter is chemical attack, which is the main mechanism. The discussion is limited to metals and limestone as indicator materials that are particularly sensitive to pollution, although many materials are used in cultural heritage objects. Later discussions present laboratory simulations to extend the simple observations and to try to answer some of the questions that have arisen. Laboratory studies permit single processes to be isolated by holding other conditions constant in ways that would be impossible in the ambient atmosphere. Other studies look at combinations of pollutants to examine synergistic effects.

The chapter then focuses on observed levels of corrosion and shows how field programmes have been used to study the process through the conceptually simple method of exposing samples of different materials to ambient atmospheres of different types. A number of programmes are presented. It also shows how earlier monitoring and field-based corrosion and soiling studies may be extrapolated to provide realistic trends for the immediate future and consideration of likely effects in areas of the world that have not had such detailed research.

Trends of corrosion are discussed based on the long-term exposures during 1987–1995 and 1997–2001 of the longstanding international collaboration

J. Tidblad (✉)
Swerea KIMAB AB, Box 55970, SE-10216, Stockholm, Sweden
e-mail: johan.tidblad@swerea.se

J. Watt et al. (eds.), *The Effects of Air Pollution on Cultural Heritage*,
DOI 10.1007/978-0-387-84893-8_3, © Springer Science+Business Media, LLC 2009

under the auspices of UNECE – the International Co-operative Programme on effect on materials including historic and cultural monuments (ICP Materials). The discussion will cover zinc, copper, bronze, medieval glass and limestone showing that corrosion has decreased during this period on average by 50%. Results from a one-year trend exposure of ICP Materials in the period 1987–2003 including carbon steel, zinc and limestone will be evaluated.

Examples of trends of steel and zinc corrosion from individual test sites with long records of corrosion rates and pollution (SO_2) e.g. Stockholm, Prague and Kopisty will be presented and discussed along with trends in corrosion of Portland limestone in London showing the "memory effect". The recent development of trends in Europe suggests that the decrease in corrosion has ceased in some regions despite the still decreasing SO_2 levels. It should also be stressed that even though the corrosion rates in industrial countries have decreased considerably the deterioration is still higher in urban areas than in surrounding rural regions.

This chapter also looks at the equations that scientists have developed to date that predict the amount of damage that will result from a given amount of pollutant. These are known as "dose-response functions" and, since they make a statistical generalisation from a number of field experiments, they give the capability of making predictions from any measured ambient pollutant levels (and meteorological parameters). This can be very powerful when we try to assess the risk of damage to other heritage buildings. A number of developments are reviewed, from the early functions developed at a time when SO_2 pollution dominated to the more recent multi-pollutant atmospheres where traffic has emerged as a dominant source. A case study that is currently looking at the situation in Asia and Africa, shows that this work remains relevant, even though pollution levels in the developed world have fallen substantially.

The fieldwork that underpins the development of the functions is labour intensive and can take many years and is therefore prohibitively expensive. Dose-response functions, therefore, are only available for a limited number of materials. Interpretation of effects on other materials is in reality related to differences in all of the many things that make materials different. Two building stones, for example, may have different mineral make ups, different physical properties such as porosity or jointing and different chemical reactivity. A practical solution is the use of some materials as indicators for the corrosion risk of other related metals or stones, which will be developed further in Chapter 8.

A final case study undertaken in the United States on Hiker Statues shows how dose-response functions can be used for predicting damage levels and discusses the implications for the conservation of bronze sculptures.

3.2 Corrosion Processes

Atmospheric corrosion is a complex field. Degradation processes can be physical, chemical, biological or a combination thereof. The effect of pollution is, however, mainly associated with a chemical attack, which is the focus

Fig. 3.1 Corroded roof of corrugated galvanised steel, the *dark* areas show the spread of rust

Fig. 3.2 Portal figures of sandstone in the Old Town in Stockholm

of the present treatment. Cultural heritage uses many materials, but this discussion is limited to metals and limestone as indicator materials particularly sensitive to pollution. Corrosion attack is mostly a non-desirable effect that causes a loss of aesthetic value and mechanical strength (Figs. 3.1 and 3.2). The corrosion products may, however, in some cases be considered beautiful – as in the case of the green patina roofs.

3.2.1 General Description of Degradation of Materials Due to Atmospheric Corrosion

One of the reasons for the complexity of the degradation process is the many phases involved. This is illustrated schematically in Fig. 3.3. In almost all cases, water is a necessary requirement for corrosion to occur. The water is either adsorbed on the metal as a thin liquid film or absorbed in the more or less porous stone material or corrosion products. Many parameters affect the degradation process. Usually they act together in different combinations and in the following, the most important parameter combinations will be described in more detail:

- The effect of SO_2 in combination with NO_2/O_3 and temperature/relative humidity
- The effect of HNO_3 in combination with temperature/relative humidity
- The effect of precipitation and acid rain
- The effect of particulate matter including NaCl in combination with temperature/relative humidity

The water film, in reality thinner than depicted in the figure, provides the link to the atmosphere in which corrosive species can dissolve and attack directly, or indirectly by modifying the properties of the water layer.

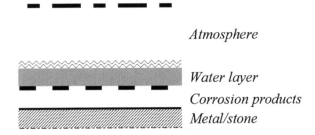

Atmosphere

Water layer

Fig. 3.3 Involved phases in the atmospheric corrosion process

Corrosion products

Metal/stone

3.2.2 The Effect of SO₂ in Combination with NO₂/O₃ and Temperature/Relative Humidity and Other Synergistic Effects

The detrimental effect of SO_2 on materials has been known since the beginning of the 20th century. The gas is dissolved into the water layer as sulphite and with the help of an oxidiser (e.g. NO_2/O_3) it is then converted to sulphate:

$$SO_2 + H_2O \rightarrow H^+ + HSO_3^- + O_3 \rightarrow 2H^+ + SO_4^{2-} + O_2 \qquad (3.1)$$

As is seen from the reaction SO_2 is an acidifying pollutant and the acidification of the water layer in turn accelerates the corrosion process. In ambient atmospheres, however, the levels of SO_2 are usually small in comparison to the levels of the oxidation agents, which therefore determine the corrosion rate.

Synergy is the interaction of two or more agents or forces so that their combined effect is greater than the sum of their individual effects. The term synergistic effect implies that the corrosion attack for a material exposed to a mixture of pollutants is greater than the sum from individual exposures acting alone.

Synergistic effects of pollutant mixtures have been unambiguously demonstrated in laboratory tests (Fig. 3.4), where single and combined corrosion effects can be simulated and controlled. Interpretation of the results of outdoor field exposures is more complicated. The variability and complexity of the outdoor atmosphere constitute a major problem for correlating the corrosion rate of a material to the concentration of a particular trace pollution component in the environment. Another potential synergistic effect, not proven at this stage in the field, is the combined effect of particulate matter and gaseous pollutants.

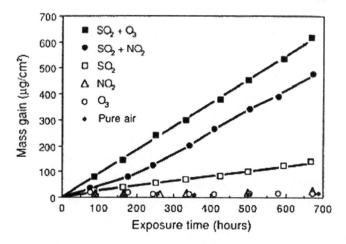

Fig. 3.4 Mass gain of zinc samples exposed in air at 95% RH. (adapted from Svensson and Johansson, 1993)

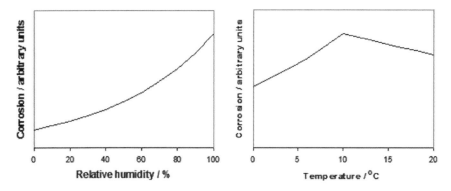

Fig. 3.5 Schematic illustration of effect of relative humidity (*left*) and temperature (*right*) for zinc expressed as annual averages in SO_2 polluted atmospheres

The dry deposition of SO_2 is influenced by both temperature and relative humidity since they are the main factors that determine the thickness of the moisture layer and therefore its ability to dissolve gases. During a day, the cyclic variation of temperature and relative humidity results in a substantial variation of the amount of water and the variation in corrosion rate can be several orders of magnitude. What are important in the long-term perspective, however, are the annual average values of temperature and relative humidity. Figure 3.5 illustrates these effects.

Corrosion increases dramatically with relative humidity and this is related to the amount of absorbed water. At low relative humidity, the water thickness is less than two monolayers and is tightly bound to the surface. At higher relative humidity, the water thickness is in the range of ten or more monolayers and is more loosely bound enabling it to act as an ionic medium for dissolving pollutants.

The effect of temperature is more complicated and for several materials a maximum is observed at about 10°C. At annual temperatures below 10°C corrosion increases with temperature and this can be related to the increased time of wetness, defined as the time when relative humidity is higher than 80% and temperature simultaneously above 0°C. The decreasing part above 10°C is attributed to a faster evaporation of moisture layers e.g. after rain or dew periods and a surface temperature above the ambient temperature due to sun radiation which results in a decrease of the surface time of wetness.

3.2.3 The Effect of HNO₃ in Combination with Temperature/Relative Humidity

The role of nitric acid (HNO_3) on the atmospheric corrosion of metals has so far received little or no attention. However, the last decades of decreasing sulphur dioxide (SO_2) levels and unchanged HNO_3 levels in many industrialised

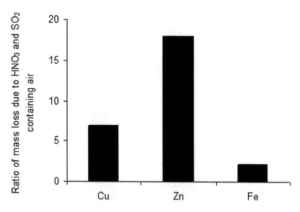

Fig. 3.6 Ratio of HNO$_3$ effect and SO$_2$ effect at comparable concentrations for copper, zinc and carbon steel

countries have resulted in an increased interest in possible HNO$_3$-induced atmospheric corrosion effects. Recent results indicate that the corrosion effect of HNO$_3$ by far exceeds that of SO$_2$, with a factor between 2 and 20 depending on the material (Fig. 3.6).

The pollution level of HNO$_3$ is typically around 1–2 µg m^{-3} (annual average) and is generally below the SO$_2$ concentration. However, due to the aggressiveness of this pollutant compared to SO$_2$ the effects of SO$_2$ and HNO$_3$ can be comparable in the present multi-pollutant situation. In the field, empirical data has proved HNO$_3$ to have an effect for zinc and limestone but so far not for copper and carbon steel.

3.2.4 The Effect of Precipitation and Acid Rain

When rain starts to fall the amount of water on the surface becomes so large that water may dissolve aggressive substances such as chlorides and transport them away from the surface, i.e. rain has a washing effect that may actually decrease the attack. In addition, it also has a corrosive effect, by wetting the surface and dissolving protective layers depending on the pH of the precipitation (acid rain effect). These two effects act in opposite directions and the net result can be either positive or negative, depending on the type of environment and material. This is illustrated in Fig. 3.7 where carbon steel and zinc have been exposed in "rural", "urban", "industrial" and "marine environments". For the rural and urban environments, the corrosion attack in unsheltered position is higher than that obtained in sheltered position, especially for zinc and the "acid rain" effect is dominating. In marine environments, it is always beneficial to remove the very aggressive chlorides and the "washing" effect dominates. In the industrial atmosphere, the result depends on the material and the relative solubility and composition of the respective corrosion products formed on carbon steel and zinc in this atmosphere.

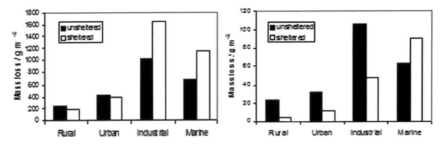

Fig. 3.7 Corrosion of carbon steel (*left*) and zinc (*right*) after 5 years of exposure

For statistical reasons it is very difficult to quantify these two opposing effects in field investigations and therefore recent dose-response functions for urban and rural areas mainly include the net effect, expressed as the effect of acid rain.

Laboratory investigations have shown that the effect of rain increases with increasing temperature but this has not been proven in field investigations.

3.2.5 The Effect of Particulate Matter Including NaCl in Combination with Temperature/Relative Humidity

Particulate matter in general is hygroscopic and starts to attract substantial amounts of water at a relative humidity below 100%. For sodium chloride, this critical relative humidity is about 75%. Chlorides are in themselves also very corrosive, as illustrated in Fig. 3.7. When combined with the effect of temperature very high corrosion rates are obtained. This is in contrast to SO_2 (Fig. 3.5) where a decrease of corrosion at temperatures above 10°C is observed. Relative humidity and temperature dependences are illustrated in Fig. 3.8. The most corrosive places in the world are thus found in warm coastal areas.

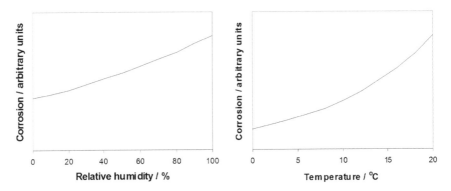

Fig. 3.8 Schematic illustration of effect of relative humidity (*left*) and temperature (*right*) for zinc expressed as annual averages in chloride containing atmospheres

Ammonium salts such as NH_4NO_3, $(NH_4)_2SO_4$ and NH_4HSO_4 are also important due to their hygroscopic properties. The ionic content of the particles prolongs the time of wetness also by affecting the freezing point thereby enabling the formation of moisture levels below $0°C$. Particles may, however, also reduce the corrosion rate by neutralising the effect of acidifying pollutants if they are basic.

3.3 Observed Levels of Corrosion

Both practical experience and systematic exposure programmes have shown that the rate of atmospheric corrosion varies greatly at different locations. It is a function of both the material and the meteorological and pollution parameters at the location.

As early as the mid-1800s it had become apparent that several materials used in objects of cultural heritage like building stones, bronze, copper and iron are sensitive to air pollution (Brimblecombe, 1987). An investigation of samples from 10 bronze monuments in Germany in 1864 to find out the reason for different patina colours revealed the influence of industrial pollution and large scale use of brown coal for heating (Magnus, 1864). In the late 1800s different deterioration rates of marble were identified at graveyards in rural and urban locations. Marble corrosion in cities was found to be "an exaggeration of the normal rate" which was demonstrated especially in positions less screened from rain (Geike, 1880). Later in the 20th century more systematic exposures were started but usually without any, or very little, environmental characterisation of the sites. It was in the beginning of the 1940s when systematic exposures of zinc and steel and simultaneous measurements of deposition of SO_2 in Berlin gave the first quantitative data on the effects on corrosion rates (Schikorr, 1943). After World War II exposures were performed in several countries-though in restricted geographical regions of single countries. Several international expo- sure programmes have been executed in different regions in the last three decades and some are described below.

3.4 Overview of Major Exposure Programs

During the last decades, a number of extensive international exposure programs have been performed or started which cover different parts of the world, Table 3.1. In the years 1980–82 the NATO/CCMS Pilot Study included exposure of sand- stone and limestone specimens and measurements of deposition of pollutants at 25 sites in Europe and two sites in the United States (Zallmanzig, 1985). The programme showed high deterioration rates in several urban and industrial loca- tions and the dominating effect of SO_2 levels.

Table 3.1 Overview of pollution parameters measured in recent major field exposures

Dry deposition							Wet deposition
Program	SO$_2$	NO$_2$	O$_3$	HNO$_3$	Cl	Particles	Precipitation (amount, composition)
NATO - CCMS	X	X			X		
ISO CORRAG	X				X		
MICAT	X				X		
ICP Materials	X	X	X				X
MULTI-ASSESS	X	X	X	X		X	X
RAPIDC	X	X	X	X		X	X

The ISO CORRAG program was initiated in 1987–1989 in order to provide corrosion and environmental data complying with ISO/TC 156 testing methods and procedures. It includes 53 test sites in 14 countries located in Europe, Argentina, Canada, Japan, New Zealand, and United States (Knotkova, 1993).

The MICAT project aimed at mapping the atmospheric corrosivity was performed at 46 test sites covering a wide range of environmental conditions in 12 Ibero-American countries and in Spain and Portugal (Morcillo et al., 1998).

A weak point of all these programmes is that the characterisation of the pollution situation is incomplete. The ISO CORRAG and MICAT programs lack data on gases other than SO$_2$ and on wet deposition. The ICP Materials programme, which aims at investigation of the effect of acid deposition on materials and is performed on non-marine sites, lacks data on dry deposition of chlorides, and also HNO$_3$ and particles. The MULTI-ASSESS programme within the European Commission 5th Framework Program performed at the test sites of the ICP Materials multi-pollutant programme and the RAPIDC project performed at 18 sites in South and South-East Asia, and Southern Africa adds in measurements of HNO$_3$ and particles. The latter programmes are therefore by far the most suitable for investigating multi-pollutant effects in non-marine environments.

3.4.1 UNECE ICP Materials Programme

The UNECE International co-operative programme on effects on materials including historic and cultural monuments (ICP Materials) is at present the most suitable for illustrating the changes that have occurred in the last decades. A short summary of the programme is given here but the interested reader should consult the results presented at the workshop on "Quantification of Effects of Air Pollutants on Materials", held in Berlin, Germany in 1998, which also includes an overview of the programme (Fitz, 1999). Updated information about the programme is also available at the ICP Materials' web page (www.corr-institute.se/ICP-Materials/).

ICP Materials is one of several effect oriented International Co-operative Programmes (ICPs) within the United Nations Economic Commission for Europe (UNECE) and the Convention on Long-range Transboundary Air Pollution (CLRTAP). The main aim of the programme is to determine the effect of acidifying air pollutants on materials used in technical constructions and in cultural heritage objects and to assess the trends of pollution and deterioration of materials. The primary objective is to collect information on corrosion and environmental data in order to evaluate dose/response functions and trend effects. This is achieved by exposing material specimens in a network of test sites (Fig. 3.9), by measuring gaseous pollutants, precipitation and climate parameters at or near each test site and by evaluating the corrosion effects on the materials.

A Task Force organises the programme with Sweden as lead country sharing the chairmanship with Italy (ENEA), with Swerea KIMAB AB (former Swedish Corrosion Institute) serving as the main research centre. Sub-centres in different countries have been appointed, each responsible for their own group of materials. The environmental sub-centre, the Norwegian Institute for Air

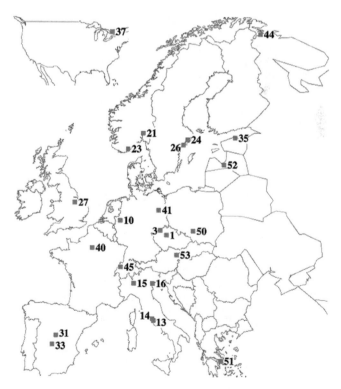

Fig. 3.9 Map of ICP Materials test sites with inset of North America (top/left)

Research (NILU) is responsible for the database of environmental parameters. In each country a national contact person has been appointed responsible for the sub-centre and/or test sites. The original network, used during the period 1987–1995, consisted of 39 exposure sites in 12 European countries and in the United States and Canada, Table 3.2. The second multi-pollutant programme started in 1997 and consisted of 29 exposure sites in 14 European countries and in Israel, the United States and Canada. At present one-year trend exposures are repeated every third year, the last completed was performed in 2005/2006 and the present on-going started in autumn 2008.

Table 3.2 List of all ICP Materials test sites including those who are currently not in use, showing number (label) of each test site, together with their name and country

No	Name	Country
1	Prague-Letnany	Czech Republic
2	Kasperske Hory	Czech Republic
3	Kopisty	Czech Republic
4	Espoo	Finland
5	Ähtäri	Finland
6	Helsinki-Vallila	Finland
7	Waldhof-Langenbrügge	Germany
8	Aschaffenburg	Germany
9	Langenfeld-Reusrath	Germany
10	Bottrop	Germany
11	Essen-Leithe	Germany
12	Garmisch-Partenkirchen	Germany
13	Rome	Italy
14	Casaccia	Italy
15	Milan	Italy
16	Venice	Italy
17	Vlaardingen	Netherlands
18	Eibergen	Netherlands
19	Vredepeel	Netherlands
20	Wijnandsrade	Netherlands
21	Oslo	Norway
22	Borregard	Norway
23	Birkenes	Norway
24	Stockholm South	Sweden
25	Stockholm Centre	Sweden
26	Aspvreten	Sweden
27	Lincoln Cathedral	United Kingdom
28	Wells Cathedral	United Kingdom
29	Clatteringshaws Loch	United Kingdom
30	Stoke Orchard	United Kingdom
31	Madrid	Spain
32	Bilbao	Spain
33	Toledo	Spain

Table 3.2 (continued)

No	Name	Country
34	Moscow	Russian Federation
35	Lahemaa	Estonia
36	Lisbon	Portugal
37	Dorset	Canada
38	Research Triangle Park	USA
39	Steubenville	USA
40	Paris	France
41	Berlin	Germany
43	Tel Aviv	Israel
44	Svanvik	Norway
45	Chaumont	Switzerland
46	London	United Kingdom
47	Los Angeles	USA
49	Antwerpen	Belgium
50	Katowice	Poland
51	Athens	Greece
52	Riga	Latvia
53	Vienna	Austria
54	Sofia	Bulgaria

3.4.2 International Standards

Within the field of exposure and evaluation of specimens under atmospheric conditions and the classification of corrosivity of atmospheres, the standardisation within The International Organisation for Standardisation, Technical Committee on Corrosion of Metals and Alloys ISO/TC 156 should be mentioned primarily. It has developed standards on atmospheric corrosion testing including general requirements for performance of field tests (ISO 8565) and a standard on procedures for removal of corrosion products from corrosion test specimens by pickling (ISO 8407) and determination of mass loss of specimens after exposure. The specified procedures are designed to remove all corrosion products without significant removal of base metal. This allows an accurate determination of the mass loss of the metal, which occurred during exposure to the corrosive environment.

The ISO/TC 156 has also developed a system for classification of *outdoor atmospheric corrosivity* based on two different approaches (ISO 9223). The first method involves exposure of standard specimens of carbon steel, zinc, copper and aluminium at the site to be classified for one year (ISO 9226). The mass loss values are used to determine the corrosion rates and the corrosivity classes of the atmosphere at that location. Table 3.3 shows the corrosivity classes based on mass loss results.

The second method is an estimation of corrosivity based on environmental information. If it is not possible to determine the corrosivity categories by

Table 3.3 ISO 9223 Corrosivity categories for carbon steel zinc, copper and aluminium based on corrosion rates

Corrosivity	Category	C-steel $\mu m\ yr^{-1}$	Zinc $\mu m\ yr^{-1}$	Copper $\mu m\ yr^{-1}$	Aluminium $g\ m^{-2}\ yr^{-1}$
Very low	C1	≤ 1.3	≤ 0.1	≤ 0.1	negligible
Low	C2	1.3–25	0.1–0.7	0.1–0.6	≤ 0.6
Medium	C3	25–50	0.7–2.1	0.6–1.3	0.6–2
High	C4	50–80	2.1–4.2	1.3–2.8	2–5
Very high	C5	80–200	4.2–8.4	2.8–5.6	5–10

exposure of standard specimens, an estimation of corrosivity can be based on corrosion loss after one year of exposure calculated from environmental data or on information on environmental conditions and exposure situation. ISO 9223–9226 is presently under revision and in the next version it is planned to give dose- response functions for the four standard metals describing the corrosion attack after one year of exposure in open air as a function of SO_2 dry deposition, chloride dry deposition, temperature and relative humidity. The functions are based on results of large field exposures and cover different climatic conditions and pollution situations. The revised version of ISO 9223 and 9224 will also include a corrosivity category covering extremely corrosive atmospheres found for instance in warm marine climates.

The corrosion rate of metals and alloys exposed to natural outdoor atmospheres is not constant with exposure time. For most metals and alloys it decreases with exposure time because of the accumulation of corrosion products on the surface of the metal exposed. Based on the dose-response functions planned to be included in the next version of ISO 9923 (Mikhailov et al., 2004) it is possible to specify guiding values of corrosion attack for metals and alloys exposed to natural outdoor atmospheres for exposures longer than one year, Table 3.4. Guiding values of corrosion damage can be used to predict the extent of corrosion damage in long-term exposures based on measurements of corrosion damage in one-year exposure to the outdoor atmosphere in question. These values may also be used to determine conservative estimates of corrosion damage based on environmental information or corrosivity category estimates as shown in ISO 9223.

For *indoor atmospheres* with low corrosivity e.g. places where works of art and historical objects are stored or electronic devices and sophisticated technical products are used, the classification in ISO 9223 is too broad. For such purposes it is necessary to subdivide the corrosivity categories C1 (very low) and C2 (low) into indoor corrosivity categories given in ISO 11844-1.The evaluation of low corrosivity indoor atmospheres can be accomplished by direct determination of corrosion attack of selected metals ISO 11844-2 or by measurement of environmental parameters ISO 11844-3 which may cause corrosion on metals and alloys. The metals used for characterisation of corrosivity of indoor atmospheres are carbon steel, zinc, copper and silver. The mass loss or mass gain of these metals is determined after one year of exposure in the atmosphere in question.

Table 3.4 Guiding corrosion values for corrosion rates (r_{av}, r_{lin}) of carbon steel, zinc and copper in atmospheres of various corrosivity categories. Values in µm per year

Metal	Average corrosion rate (r_{av}) during the first 10 years for the following corrosivity categories					
	C1	C2	C3	C4	C5	CX
Carbon steel	$r_{av} \leq$ 0.4	$0.4 < r_{av}$ ≤ 8.3	$8.3 < r_{av}$ ≤ 17	$17 < r_{av}$ ≤ 27	$27 < r_{av}$ ≤ 67	$67 < r_{av}$ ≤ 233
Zinc	$r_{av} \leq$ 0.07	$0.07 < r_{av}$ ≤ 0.5	$0.5 < r_{av} \leq$ 1.4	$1.4 < r_{av}$ ≤ 2.7	$2.7 < r_{av}$ ≤ 5.5	$5.5 < r_{av}$ ≤ 16
Copper	$r_{av} \leq$ 0.05	$0.05 < r_{av}$ ≤ 0.3	$0.3 < r_{av} \leq$ 0.6	$0.6 < r_{av}$ ≤ 1.3	$1.3 < r_{av}$ ≤ 2.6	$2.6 < r_{av}$ ≤ 4.6

Metal	Steady state corrosion rate (r_{lin}) estimated as the average corrosion rate during the first 30 years for the following corrosivity categories					
	C1	C2	C3	C4	C5	CX
Carbon steel	$r_{lin} \leq$ 0.3	$0.3 < r_{lin}$ ≤ 4.9	$4.9 < r_{lin}$ ≤ 10	$10 < r_{lin}$ ≤ 16	$16 < r_{lin}$ ≤ 39	$39 < r_{lin}$ ≤ 138
Zinc	$r_{lin} \leq$ 0.05	$0.05 < r_{lin}$ ≤ 0.4	$0.4 < r_{lin}$ ≤ 1.1	$1.1 < r_{lin}$ ≤ 2.2	$2.2 < r_{lin}$ ≤ 4.4	$4.4 < r_{lin}$ ≤ 13
Copper	$r_{lin} \leq$ 0.03	$0.03 < r_{lin}$ ≤ 0.2	$0.2 < r_{lin}$ ≤ 0.4	$0.4 < r_{lin}$ ≤ 0.9	$0.9 < r_{lin}$ ≤ 1.8	$1.8 < r_{lin}$ ≤ 3.2

As a scientific approach is essential for the conservation of cultural property a specific European standardisation activity has been recently started within the CEN/TC 346 Conservation of Cultural Property. The main objective of this technical committee is drafting European standards on well-experimented test methods both in laboratory and in situ which will help conservation professionals in their restoration and conservation work. This standardisation activity will permit a harmonisation and unification of methodologies for the whole European area. The following fields will be the subject of the standardisation activity:

- terminology relevant to movable and immovable artefacts and to their conservation and to materials constituting the artefacts
- guidelines for a methodological approach to the knowledge of artefacts and conditions of optimum long-term preservative conservation
- test and analysis methods for the diagnosis and for the characterisation of the artefacts and of their state of conservation, for the evaluation of the performance of products and methodologies for conservation work and for the evaluation of conservation conditions of indoor cultural property.

3.4.3 Typical Corrosion Rates for Different Materials

Even if the degradation process described in this chapter has a general validity the degradation mechanism is specifically dependent on the material. Corrosion rates differ for individual materials. This applies both for the effect of different pollutants and for the development of the corrosion rate with time.

There are thus great differences in corrosion rates in different types of atmospheres from rural to urban/industrial. Corrosion in marine atmospheres is affected by the strong effect of chlorides from marine aerosols. The variation of corrosion rates can be illustrated by results from the ICP Materials exposure programme at test sites in Europe and North America, Fig. 3.10. The results for bronze and Portland limestone show the wide spread of corrosion rates from

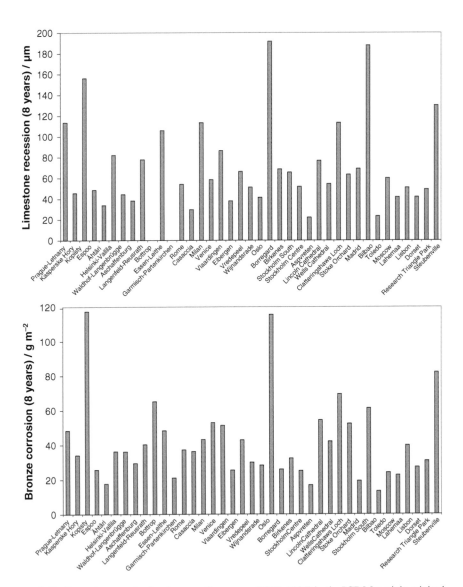

Fig. 3.10 Corrosion attack after 8 years of exposure (1987–1995) in the ICP Materials original exposure programme of Portland limestone (*top*) and cast bronze (*bottom*)

very low rates at background rural stations like Aspvreten and Ahtari in Scandinavia and the industrial sites Kopisty in Czech Republic, Borregard in Norway and Steubenville in the United States.

There are, however, also differences in corrosion rates in microclimates which may be very important for planning maintenance on objects of cultural heritage. This is demonstrated by results from exposure at the Royal Palace in Stockholm and in the Prague castle showing the differences in corrosion rate in different positions on the building, Fig. 3.11. The figure also demonstrates the big difference in the level of pollution in the two cities. In Stockholm measures taken since the late 1960s comprising a reduction of sulphur content in oil products and an expansion of the district heating system has lowered the SO_2 concentration to rural levels in the first half of the 1990s. Pollution in Prague at this time and consequently the corrosion rates were much higher. Since then even in Prague SO_2 levels have decreased to around 10 μgm^{-3} with a subsequent reduction in corrosion rate.

It should also be emphasized that during an initial period the corrosion of some materials like steel and weathering steel, and to a lesser extent zinc and copper, is usually higher before the protective properties of rust and other

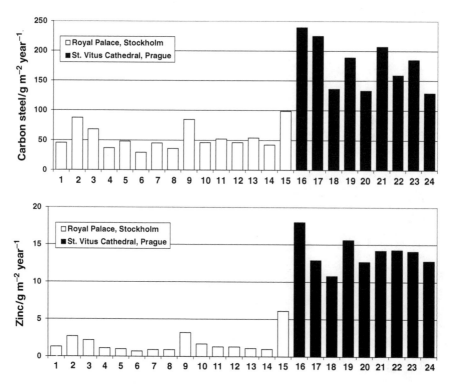

Fig. 3.11 Corrosion rate of carbon steel (*top*) and zinc (*bottom*) after 1 year of exposure in Stockholm and Prague

Table 3.5 Steady state corrosion rates in different types of atmosphere, μm yr^{-1}

Type of material	Rural atmosphere	Urban atmosphere
SO$_2$-conc. μg m^{-3}	*<10*	*10–100*
C-steel	5–10	10–30
Weathering steel	2–5	2–6
Zinc	0.2–2	2–5
Copper	0.2–0.6	0.6–2
Bronze	0.2–0.4	0.4–2
Portland limestone	3–6	6–40

corrosion products improve and the corrosion rate decreases and reach "steady state" values. Steady state corrosion rates of some important materials are given in Table 3.5 for rural and urban atmospheres. Further data on maximum corrosion rates after prolonged exposure periods are planned to be included in the ISO 9224 revised standard. These types of values are useful for decision on material choices and for planning of maintenance or restoration of objects of cultural heritage, where the expected lifetimes are usually long.

3.4.4 Long-Term Trends in the 20th Century

The first outdoor field exposure programmes started in the beginning of the 20th century but the environmental characterisation was limited to verbal descriptions such as "Rural", "Marine", "Urban" and "Industrial". One of the longest trend series of parallel measurements of corrosion and SO$_2$ pollution has been collected in Stockholm, Sweden as is illustrated in Fig. 3.12 for zinc. A series of events has contributed to the changes in the corrosion rate. Before 1945, individual domestic heating by burning wood was prevailing. In the beginning of the 1950s, oil with high S content gradually replaced the earlier used fuels. Therefore, both the SO$_2$ concentration and the corrosion rate rose dramatically. In 1959, the first district heating facility was started with a continued extension since then. In 1966, when the SO$_2$ concentration peaked, the S content of oil was on average 1.6%. It has then been lowered in several steps, with accompanying decreases in corrosion, as a result of legislative change.

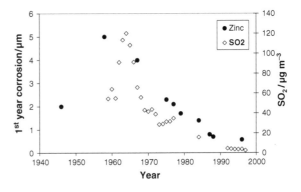

Fig. 3.12 First year corrosion attack of zinc (fresh samples) and annual averages of SO$_2$ concentration measured in Stockholm, Sweden (site "Vanadis")

The industrial site, Kopisty, is located about 90 km northwest of Prague. In 1994, a program of desulphurization in all power plants was started and the total invested sum has been 46 billion Czech crowns. Also, the changes in industrial structure and technologies have resulted in a decrease of the total power consumption of about 20%. The trends of carbon steel corrosion and SO_2 pollution gathered at Kopisty are shown in Fig. 3.13.

The similarity between Figs. 3.12 and 3.13 is striking with a coinciding peak of SO_2 and corrosion illustrating the strong effect of SO_2 on corrosion as well as the large benefits achieved with lowered pollution levels. The difference is in the time of the peak, occurring earlier in Scandinavia and later in central Europe.

Mid-20th century trends in zinc corrosion rates parallel the improvement in air quality in the United States. Zinc corrosion decreased in small cities, and more dramatically so in large industrial areas between the 1930s–1980s. Figure 3.14

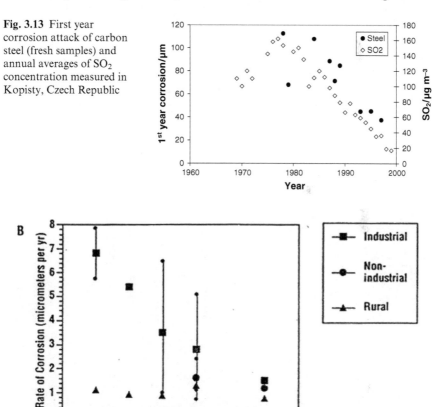

Fig. 3.13 First year corrosion attack of carbon steel (fresh samples) and annual averages of SO_2 concentration measured in Kopisty, Czech Republic

Fig. 3.14 Zinc corrosion rates in North America 1931–1988. The data points are averages of all available corrosion data for longer than one year exposure. Small dots represent minimum and maximum values for the decade.

compares estimates of historical SO_2 levels (based on fuel use and demographics) with American Society for Testing Materials long-term zinc corrosion measurements. Corrosion rates for zinc have decreased since the 1930s, when the earliest measurements of ambient corrosion were made in the United States. This improvement in metal performance is generally associated with decreases in sulphur pollutant concentrations in industrial and urban areas over the same period. While these estimates of long-term pollution exposure may suffer from lack of precision or the absence of confirming environmental measurements in the early years, their ability to predict zinc corrosion is remarkable.

The one-year exposure of the original ICP Materials programme (1987–88) and the multi-pollutant exposure programme (1997–98) constitute a pair of one-year exposures that can be used to evaluate trend effects for all the materials included in both of the exposures. This is the period in time with large reductions in both pollution and corrosion (Figs. 3.12 and 3.13). The result of the comparison is shown in Fig. 3.15. The figure shows an average trend for all ICP Materials test sites but it is worth noting that the decrease is present at both high (urban/industrial) and low (rural) polluted sites with corresponding high and low corrosion attack. On average, the decrease in both corrosion and pollution is about 50% for this 10-year period. However, marked exceptions are zinc and SO_2, with a larger decrease compared to the other materials/pollutants and copper and O_3, with a smaller or nonexistent decrease. Note that copper is the only inorganic material where O_3 is included in the dose-response function (see below).

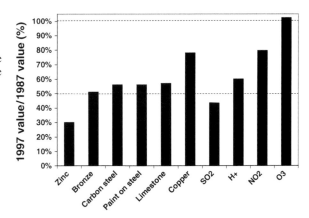

Fig. 3.15 Ratio between first year corrosion attacks and annual averages of pollutants obtained from the ICP Materials first exposure (1987/88) vs. the multi-pollutant exposure (1997/98). The ratios are based on averages of individual urban, rural and industrial sites

3.4.5 Recent Trends

For several sites, especially in northern Europe, the decrease in corrosion attack has not continued in the most recent years and the levels of corrosion attack are now constant with time for both rural as well as urban sites. The break in the trend, from decreasing to constant, observed for the sites in Scandinavia has occurred or will occur later for other sites as is illustrated in Fig. 3.16 for carbon steel.

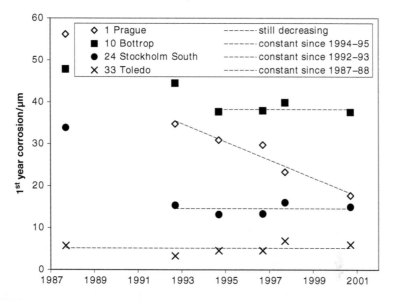

Fig. 3.16 Corrosion attack of unsheltered carbon steel after 1 year of exposure (μm)

If comparing different materials, it can be said that during the period 1987–1997 the decreasing trend in the concentrations of acidifying air pollutants resulted in a mirrored decrease in corrosion of carbon steel, zinc and limestone. During the period 1997–2003, however, the corrosion rate of carbon steel decreased while the corrosion rate of zinc and limestone increased slightly. The most recent one-year exposure of carbon steel, zinc and limestone was performed during 2005–2006 and now also for carbon steel, the average trend is no longer decreasing Fig. 3.17.

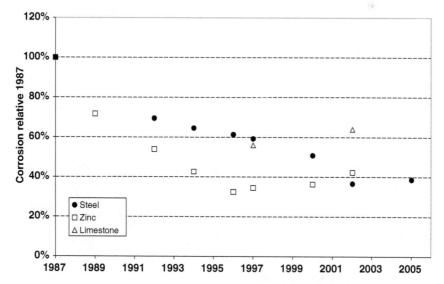

Fig. 3.17 Average trends in corrosion relative to the first exposure in 1987

3.5 Dynamic Effects (Memory Effects)

Despite significant reductions in atmospheric sulphur dioxide levels, it has been observed in several cases that building facades still degraded at an apparently unchanged rate. This is the case for weathering observed on natural stone buildings and the relationship between the reductions in atmospheric pollution and reductions in the rate of weathering. It is in contrast to the corrosion of metals where a reduction in SO_2 has an immediate effect on the reduction in corrosion.

The preferred explanation, the so called "memory effect", attributes the faster weathering of old stone to its past exposure history. Studies carried out directly on the stone surfaces at St. Paul's Cathedral in London between 1980 and 2000 using the so called "micro-erosion meter" show this clearly (Fig. 3.18). St. Paul's Cathedral is in central London in the heart of an area that was heavily polluted for more than 200 years. In the same period the atmospheric SO_2 concentration had continued to decline but the rate of change in the weathering appears to have a "time-lag" in the order of 15–20 years, illustrated schematically in Fig. 3.19. It should be emphasised that even the last values from the cathedral surfaces are about four times the value of the recession of fresh stone specimens from the period 1997–2001.

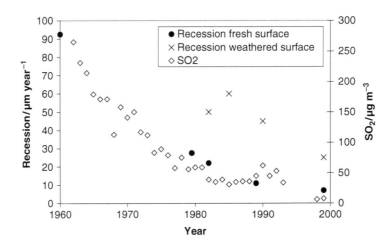

Fig. 3.18 Degradation of samples of fresh Portland limestone exposed in Central London and of weathered surfaces of St Paul's Cathedral in London measured by "micro-erosion readings" and the SO_2 concentrations in the period 1960–2000 (From Kucera, Tidblad and Yates, 2004)

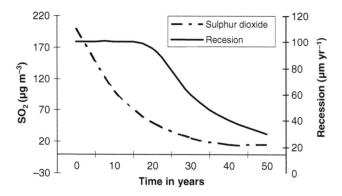

Fig. 3.19 Illustration of the memory effect for natural stone buildings where the decline in stone degradation has a time lag of about 15–20 years after the decrease of SO_2 concentration

3.6 Dose-Response Functions for Corrosion

3.6.1 Discussion of Terms

Emissions of pollutants or other chemical, physical and biological effects lead first to impacts. It is only if people consider these impacts as undesirable and put a value on them that they can be referred to as damages. A dose of a pollutant depends on both the concentration and the time of exposure and in a strict sense the dose is defined as the quantity of a pollutant that is actually delivered to the receptor. However, the term "dose-response function" is often used in a broader sense where the dose is replaced by the concentration, implicitly assuming that there is a direct relation between the concentration and deposition of the pollutant. For these types of functions, also the term exposure-response function is frequently used.

In the workshop on "Economic Evaluation of Air Pollution Abatement and Damage to Buildings including Cultural Heritage", held in Stockholm, Sweden January 23–25, 1996, the following definitions were adopted:

A *dose-response function* links the dose of pollution, measured in ambient concentration and/or deposition, to the rate of material corrosion;

A *physical damage function* links the rate of material corrosion (due to the pollution exposure given by the dose-response function) to the time of replacement or maintenance of the material. Performance requirements determine the point at which replacement or maintenance is considered to become necessary;

A *cost function* links changes in the time of replacement, repair or repainting to monetary cost; and

An *economic damage function* links cost to the dose of pollution, as derived from the three definitions above.

The term "damage function" is also frequently used to denote a function that links the degradation to other parameters that cannot be expressed as doses, for example, freeze-thaw cycles causing degradation of stone materials without involving the time of replacement/maintenance. This is not consistent with the second definition above and the term "impact function" would perhaps be more appropriate.

3.6.2 Early Dose-Response Functions

The first systematic field exposures involving simultaneous measurements of corrosion attack and SO_2 pollutant concentration was, to our knowledge, performed by Schikorr and Schikorr (1943) during the period 1939 to 1942 in Berlin, Germany (!). They spoke about "äquivalenz zwischen schwefelwert und zinkangriff" (equivalence between sulphur value and zinc attack), which based on their paper, can be quantified as

$$R_{Zn} = 0.044[SO_2]$$

where R_{Zn} is the corrosion attack in μm $year^{-1}$ and $[SO_2]$ is the SO_2 concentration in μg m^{-3}. Worth noting is that their measured SO_2 concentrations were in the range 20–500 μg m^{-3}. Although their data was only based on monthly experiments the equation is very similar to a series of functions that were developed in the following 30–40 years, all of them involving SO_2 as the main explanatory factor. Other parameters were also used in the early functions but were generally limited to acidity and of course chloride. There was also usually a wetness term and sometimes a temperature term.

3.6.3 Dose-Response Functions for the SO_2-Dominating Situation

Recent exposure programmes, including ICP Materials, have been described in an earlier section of this chapter. This programme was innovative in combining a broad characterisation of the environment, a large number of test sites and a wide selection of exposed materials. The most ambitious exposure programme within ICP Materials was performed during the period 1987–1995 and involved exposure of materials in unsheltered and sheltered positions after 1 (1987–1988), 2 (1987–1999), 4 (1987–2001) and 8 (1987–2005) years of exposure.

It was emphasised in the statistical evaluation of data that it should be based on physical and chemical principles. Therefore, for an unsheltered position the material damage was quantified in terms of dry and wet deposition of pollutants. Wet deposition included transport by means of precipitation and dry deposition transport by any other process. The dose-response relations from ICP Materials were all expressed in the general form

$$K = f_{dry}(T, Rh, [SO_2], [O_3], \ldots) + f_{wet}(Rain, [H^+])$$

where K is the corrosion attack, f_{dry} is the dry deposition term and f_{wet} is the wet deposition term. A list of all dose-response functions for exposure of unsheltered materials is shown in Table 3.6.

Table 3.6 Dose-response functions for unsheltered materials for the SO_2-dominating situation

Material	r^2	n
Structural metals		
Weathering steel (C<0.12%, Mn 0.3–0.8%, Si 0.25–0.7%, P 0.07–0.15%, S<0.04%, Cr 0.5–1.2%, Ni 0.3–0.6%, Cu 0.3–0.55%, Al<0.01%)		
$ML = 34[SO_2]^{0.13}exp\{0.020Rh + f(T)\}t^{0.33}$	0.68	148
$f(T) = 0.059(T-10)$ when $T \leq 10°C$, otherwise $-0.036(T-10)$		
Zinc		
$ML = 1.4[SO_2]^{0.22}exp\{0.018Rh + f(T)\}t^{0.85} + 0.029Rain[H^+]t$	0.84	98
$f(T) = 0.062(T-10)$ when $T \leq 10°C$, otherwise $-0.021(T-10)$		
Aluminium		
$ML = 0.0021[SO_2]^{0.23}Rh \cdot exp\{f(T)\}t^{1.2} + 0.000023Rain[Cl^-]t$	0.74	106
$f(T) = 0.031(T-10)$ when $T \leq 10°C$, otherwise $-0.061(T-10)$		
Copper		
$ML = 0.0027[SO_2]^{0.32}[O_3]^{0.79}Rh \cdot exp\{f(T)\}t^{0.78} + 0.050Rain[H^+]t^{0.89}$	0.73	95
$f(T) = 0.083(T-10)$ when $T \leq 10°C$, otherwise $-0.032(T-10)$		
Bronze (Cu $Sn_6Pb_7Zn_5$, ISO/R 1338)		
$ML = 0.026[SO_2]^{0.44}Rh \cdot exp\{f(T)\}t^{0.86}+0.029Rain[H^+]t^{0.76}$	0.81	144
$\qquad +0.00043Rain[Cl^-]t^{0.76}$		
$f(T) = 0.060(T-11)$ when $T \leq 11°C$, otherwise $-0.067(T-11)$		
Stone materials		
Limestone (Portland)		
$R = 2.7[SO_2]^{0.48}exp\{-0.018T\}t^{0.96} + 0.019Rain[H^+]t^{0.96}$	0.88	100
Sandstone (White Mansfield dolomitic sandstone)		
$R = 2.0[SO_2]^{0.52}exp\{f(T)\}t^{0.91} + 0.028Rain[H^+]t^{0.91}$	0.86	101
$f(T) = 0$ when $T \leq 10°C$, otherwise $-0.013(T-10)$		
Paint coatings		
Coil coated galvanized steel with alkyd melamine		
$L = [5/(0.084[SO_2] + 0.015Rh + f(T) + 0.00082Rain)]^{1/0.43}$	0.73	138
$f(T) = 0.040(T-10)$ when $T \leq 10°C$, otherwise $-0.064(T-10)$		
Steel panels with alkyd		
$L = [5/(0.033[SO_2] + 0.013Rh + f(T) + 0.0013Rain)]^{1/0.41}$	0.68	139
$f(T) = 0.015(T-11)$ when $T \leq 11°C$, otherwise $-0.15(T-11)$		

where
ML = mass loss, g m^{-2}
R = surface recession, μm
t = exposure time, years
L = maintenance interval (life time), years
Rh = relative humidity, % – annual average
T = temperature, °C – annual average
$[SO_2]$ = concentration, μg m^{-3} – annual average
$[O_3]$ = concentration, μg m^{-3} – annual average
$Rain$ = amount of precipitation, m year^{-1} – annual average
$[H^+]$ = concentration, mg l^{-1} – annual average
$[Cl^-]$ = concentration, mg l^{-1} – annual average

Table 3.7 Materials, parameters and their inclusion in dose-response functions for the SO_2 dominating situation for unsheltered from the first ICP Materials exposure programme (1987–1995)

Material	T	Rh	SO_2	O_3	Rain	H^+	Cl^-
Weathering steel	X	X	X				
Zinc	X	X	X		X	X	
Aluminium	X	X	X		X		X
Copper	X	X	X	X	X	X	
Cast Bronze	X	X	X		X	X	X
Portland limestone	X		X		X	X	
White Mansfield sandstone	X		X		X	X	
Coil coated galvanised steel with alkyd melamine	X	X	X		X		
Steel panels with alkyd	X	X	X		X		
Glass M1 with composition representative of medieval stained glass windows		X	X				

The equations are valid for regions without strong influence of sea salts with a chloride content in precipitation < 5 mg l^{-1} approx.

The functions for paint coatings are expressed as lifetime equations. These lifetimes can be mapped but the functions cannot be used for calculating acceptable levels/loads using the concept of acceptable corrosion rates.

Table 3.7 shows the parameters included in the listed dose-response functions for the different materials

Even though NO_2 in gaseous form as well as NO_3^- in precipitation was measured and included in the statistical analysis it was not considered as a significant parameter for the unsheltered dose-response functions. The equations show that for most metals dry deposition of SO_2 is the most important pollutant parameter except for copper where O_3 is also important. Therefore, these set of functions are referred to as dose-response functions for the SO_2 dominating situation. Wet deposition (Rain $[H^+]$) is generally

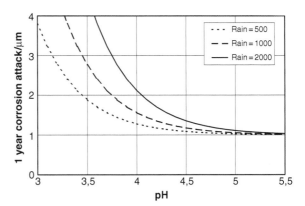

Fig. 3.20 Corrosion attack of unsheltered copper vs. pH of precipitation calculated from the dose-response function and assuming a constant dry deposition term corresponding to a corrosion attack of 1.0 μm

important for unsheltered conditions except for weathering steel and aluminium. For aluminium there is an effect of chlorides which also applies for bronze.

When the SO_2 concentration was highest, around 1960–1970 (depending on the location) the effect of wet deposition was considered small in comparison. However, as is illustrated in Fig. 3.20 for copper, the effect depends strongly on pH and amount of precipitation. At about pH 4.5 and lower, the effect cannot be neglected. Curves are shown for the precipitation (Rain) equal to 500, 1000 and 2000 mm yr^{-1}.

3.6.4 Dose-Response Functions for the Multi-pollutant Situation

As has been described previously, the decreasing levels of SO_2 and increasing levels of car traffic resulted in a multi-pollutant situation. In order to quantify this in the form of dose-response functions a new multi-pollutant exposure programme was initiated within ICP Materials. It took place from 1997–2001 and involved exposures after 1 (1997–1998), 2 (1997–1999) and 4 (1997–2001) years of exposure. It was even more ambitious than the first exposure programme described above in the characterisation of environment but included fewer test sites and less materials. Besides the parameters measured in the original programme, measurements of HNO_3 and particulate matter were included as an extension within the EU 5FP MULTI-ASSESS project. A list of all dose-response functions for exposure of unsheltered materials is shown in Table 3.8.

Table 3.8 Dose-response functions for unsheltered matrials for the multi-pollutant situation

Material
Carbon steel
$R = 6.5 + 0.178[SO_2]^{0.6}Rh_{60}e^{f(T)} + 0.166Rain[H^+] + 0.076PM_{10}$
$f(T) = 0.15(T-10)$ when $T<10°C$, $-0.054(T-10)$ otherwise
Zinc
$R = 0.49 + 0.066[SO_2]^{0.22}e^{0.018Rh+f(T)} + 0.0057Rain[H^+] + 0.192[HNO_3]$
$f(T) = 0.062(T-10)$ when $T<10°C$, $-0.021(T-10)$ otherwise
Cast Bronze
$R = 0.15 + 0.000985[SO_2]Rh_{60}e^{f(T)} + 0.00465Rain[H^+] + 0.00432PM_{10}$
$f(T) = 0.060(T-11)$ when $T<11°C$, $-0.067(T-11)$ otherwise
Portland limestone
$R = 4.0 + 0.0059[SO_2]RH_{60} + 0.054Rain[H^+] + 0.078[HNO_3]Rh_{60} + 0.0258PM_{10}$

where
Rh_{60} = Rh – 60 when Rh > 60, 0 otherwise
$[HNO_3]$ = concentration, $\mu g\ m^{-3}$ – annual average
$PM10$ = concentration, $\mu g\ m^{-3}$ – annual average
And other symbols as in table 3.6

Table 3.9 Materials, parameters and their inclusion in dose-response functions from the multi-pollutant exposure programme (1997–2003)

Material	T	Rh	SO_2	O_3	HNO_3	PM_{10}	Rain	H^+
Carbon steel	X	X	X			X	X	X
Zinc	X	X	X		X		X	X
Copper	X	X	X	X			X	X
Cast Bronze	X	X	X			X	X	X
Portland limestone		X	X		X	X	X	X

Table 3.9 shows the parameters included in the multi-pollutant dose-response functions for the different materials.

Even though SO_2 concentration has decreased substantially it is still included in all functions, as is the effect of wet acid deposition (acid rain). For copper O_3 is also important. The new parameter HNO_3 was significant for zinc and limestone while PM_{10} was included in the functions for carbon steel, cast bronze and limestone.

Dose-response functions are useful for mapping areas of increased risk of corrosion and maps are a powerful tool to illustrate the effects of pollutants on cultural heritage objects. The dose-response functions can also be used for calculation of corrosion costs and for assessing tolerable levels of pollution (see Chapter 9).

As an example, in CULT-STRAT corrosion mapping with high resolution was performed for the central European countries; Germany, Czech Republic, Austria and Switzerland (Fig. 3.21). Work that may be very valuable for higher

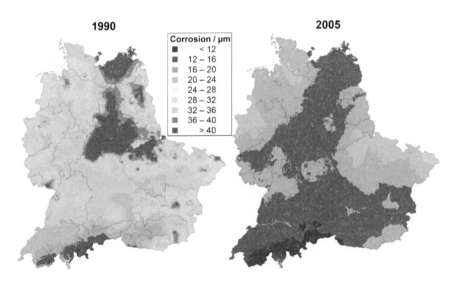

Fig. 3.21 Mapping corrosion rates of carbon steel in Germany, Czech Republic, Switzerland and Austria in 1990 and 2005 – from the CULT-STRAT final report

resolution corrosion mapping on the European level is presently going on within the ETC (European Topic Centre) on Air and Climate Change under the auspices of EEA (European Environmental Agency). The aim of ETC is to map all the pollution parameters that are also needed for the corrosion mapping in CULT-STRAT, as annual mean values at high resolution, e.g. 10×10 km. In the EU project CityDelta the results from different dispersion models used in a number of European main cities have been compared. Two typical systems for local air quality modelling and mapping are the IMMISluft – concept used by IVU Umwelt Gmbh in Germany, and the AirQuis system from The Norwegian Institute for Air Research in Norway.

In general, most of the materials including those of cultural monuments are located in urban areas. Therefore it is of great interest to produce maps showing to what extent materials in urban agglomerations are exposed to risk of increased corrosion. With the same methodology as for nationwide mapping corrosivity maps can be produced for urban areas if the necessary environmental data are available. Examples are given in Chapter 8.

3.7 Case Study: Corrosion Effects in Warm Regions of Asia and Africa

Research on the effects of acidifying pollutants on materials has mainly been performed in Europe and North America and therefore these effects are relatively well described in regions of temperate climate. Early research was focused on the effects of sulphur pollutants. While SO_2 levels in developed countries are decreasing, pollution levels due to energy production and vehicle numbers in many developing countries are rapidly increasing. In addition, many developing countries happen to be located in warm regions with high relative humidity and a high frequency of precipitation. Thus, there is a higher risk of extreme corrosion rates in these areas, compared to those with similar emission levels in the temperate zone, with accompanying high costs due to corrosion damage. The rich variety of cultural heritage objects in warm regions emphasises the need for a deeper understanding of deterioration processes. This will also help decision makers in developing countries to avoid the mistakes associated with extensive emissions of pollutants and consequent materials damage which has been the case in industrial countries for a very long time period.

3.7.1 Overview of Results of Major Programmes in Tropical and Subtropical Climates

Atmospheric corrosion research in tropical and subtropical climates started around 1945 and results are available from several individual countries; e.g. Australia, Brazil, China, Cuba, India, New Zealand, Nigeria, Panama, Papua

New Guinea, Philippine Islands, South Africa, Taiwan, Singapore and Vietnam (Tidblad et al., 2000). Ambitious programmes involving more than a few countries are, however, limited to the MICAT project (Morcillo et al., 1998) which includes 12 Ibero-American countries, Spain and Portugal, and the ISO CORRAG programme with more than 50 sites located in Europe, Argentina, Canada, Japan, New Zealand and United States (Knotkova, 1993). In Asia, the programme co-ordinated from Australia and involving 13 sites in Australia, Philippines, Thailand and Vietnam (Cole, 2000) and the programme co-ordinated from Japan and involving 23 sites in Japan, China, and South Korea (Maeda et al., 2001) should be mentioned. In Africa, the long-term study of Callaghan (1991) should not be left unnoticed. These programmes either have an environmental characterisation limited to SO_2 as a main pollutant (Morcillo et al., 1998; Knotkova, 1993; Callaghan, 1991) or have relatively short exposure times (Cole, 2000; Maeda et al., 2001).

3.7.2 RAPIDC Corrosion Programme

Partly in order to fill this gap the project "RAPIDC/Corrosion" was initiated as a part of the 2001–2004 Swedish International Development Cooperation Agency (SIDA) funded Programme on Regional Air Pollution in Developing Countries (RAPIDC). The corrosion part of the project "CORNET" is co-ordinated by Swerea KIMAB (formerly Swedish Corrosion Institute - SCI). Since this is a very extensive exposure programme with a systematic characterisation of the environmental parameters the project and the results and conclusions after one year of exposure will be presented. They will serve for the characterisation of corrosion conditions in warm regions and for a comparison with the results from temperate regions. Since one year of exposure is a relatively short period the conclusions presented are preliminary in anticipation of the results after 2 and 4 years which will be available shortly.

3.7.2.1 Test Sites and Exposure Conditions

Each site was maintained by a dedicated partner who was responsible for the safety of the rack, the exposure/withdrawal of passive samplers, the collection of environmental data and the withdrawal of corrosion specimens. After exposure, the samples were returned to Swerea KIMAB for evaluation of corrosion attack. A map and a list of test sites including the responsible organisations of the exposure are shown in Fig. 3.22 and Table 3.10. The network of test sites where the exposure started in 2002 consisted of 6 partners (12 sites) in Asia and 3 partners (4 sites) in Africa. In 2006 new sites were started in Tanzania, Mozambique, Sri Lanka, Nepal, Iran and India (Taj Mahal, Agra) and in 2008 a site has been started on the Maldives resulting in a network of 23 sites.

Africa (6 sites)   **Asia (16 sites)**

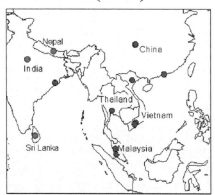

Fig. 3.22 Test sites in the CORNET exposure programme in Africa and Asia. The exposure on most marked sites started in 2002 with the exception of the eastern South African, the Tanzanian, the Nepalese, the northern Indian and the Sri Lankan site which all started in 2006

Table 3.10 List of CORNET test sites including country and responsible organisation where exposure started in 2002 and sites where exposure started in 2006 (marked with *)

Country	Test site name	Responsible organisation
India	Bhubaneswar-urban	Regional Research Laboratory
India	Bhubaneswar-rural	Regional Research Laboratory
Thailand	Bangkok	Thailand Inst. of Scientific and Technical Research
Thailand	Phrapradaeng	Thailand Inst. of Scientific and Technical Research
Vietnam	Hanoi	Institute of Materials Science
Vietnam	Ho Chi Minh City	Ho Chi Minh Branch of the Inst. of Materials Science
Vietnam	Mytho	Ho Chi Minh Branch of the Inst. of Materials Science
China	Chongqing	Chongqing Inst. of Env. Science and Monitoring
China	Tie Shan Ping	Chongqing Inst. of Env. Science and Monitoring
China	Hong Kong	Hong Kong Environmental Protection Department
Malaysia	Kuala Lumpur	Malaysian Meteorological Service
Malaysia	Tanah Rata	Malaysian Meteorological Service
South Africa	Johannesburg	CSIR/Materials and Manufacturing
Zambia	Kitwe	University of Zambia
Zambia	Magoye	University of Zambia
Zimbabwe	Harare	University of Zimbabwe
*India	Agra, Taj Mahal	Central Pollution Control Board
*Iran	Teheran	Iran Department of the Environment
*Mozambique	Maputo	Universidade Eduardo Mondlane
*Nepal	Kathmandu	Int. Centre for Integrated Mountain Development
*Sri Lanka	Colombo	Central Environmental Authority.
*Tanzania	Dar es Salaam	University of Dar es Salaam

At each site a total of nine samples were exposed in the summer of 2002 for each material and this means that there are three sets of triplicates and that in total the samples are sufficient for three exposure periods. At present sample withdrawals have been made after 1, 2 and 4 years of exposure.

The exposed metallic materials consist of panels of carbon steel, zinc, copper and painted steel with two layers of alkyd (90 μm). Portland limestone specimens of dimensions $50 \times 50 \times 8$ mm^3 were obtained from the Building Research Establishment Ltd, United Kingdom, where also the corrosion attack is evaluated as mass change during exposure. The mass change is then recalculated to surface recession.

Passive sampling was performed on all sites for the gaseous pollutants SO_2, NO_2, O_3 and HNO_3 and for particulate matter. Sampling was performed on a bi-monthly basis. The total sampling period was one year making a total of six bi-monthly sampling periods. Complementary data on temperature, relative humidity, amount of rain and its pH were collected by the partners at a nearby environmental station and were reported to Swerea KIMAB on a monthly basis.

3.7.3 Environmental and Corrosion Data

The most relevant environmental data are given in Table 3.11. Compared to the situation in Europe the values are similar except for temperature, amount of precipitation and SO_2 concentration, where the values are generally higher than the European values. The SO_2 values are generally below 20 μg m^{-3} approximately except for four extreme sites: Phrapradaeng, Chongqing, Tie Shan Ping and Kitwe. Worth noting is that the site Tie Shan Ping is a rural site situated close to Chongqing. The site Kitwe is located in the copper belt area in the northern part of Zambia. Other extreme sites worth mentioning are those in Malaysia: Kuala Lumpur and Tanah Rata. Kuala Lumpur has the highest HNO_3 values in the network probably due to a combination of the high NO_2 emissions, the high temperature, and the humid conditions. Tanah Rata is the cleanest site in the network. It has the same precipitation level as Kuala Lumpur but a much lower temperature and is situated in the Cameron highlands.

3.7.4 Comparison of Corrosion in Warm and Temperate Regions

As the methodology and materials used in the RAPIDC project are identical with the ICP Materials exposure programme the results obtained can be used for a comparison of the conditions in warm and temperate regions, Fig. 3.23.

A comparison of the results shows that while the levels of S-pollutants have decreased dramatically in industrial countries, the levels of SO_2 are high in several locations in developing countries like Phrapradaeng, Chongqing, Tie Shan Ping and Kitwe. The value of HNO_3 from Kuala Lumpur is higher than

Table 3.11 List of CORNET test sites including number, country and name and selected environmental (annual mean values) and corrosion data from the first year of exposure

No Country	Name	T	Rh	Rain	pH	HNO$_3$	SO$_2$	NO$_2$	O$_3$	Steel	Zn	Cu	Limestone
		°C	%	mm		µg m^{-3}				g m^{-2}			µm
3 India	Bhubaneswar-u	26.5	69	425	6.0	1.3	4	11	63	157	4.3	8.3	13.3
4 India	Bhubaneswar-r	26.5	69	425	6.0	1.0	3	5	63	156	3.6	11.9	9.4
5 Thailand	Bangkok	29.3	76	1371	6.8	2.3	11	39	38	116	4.6	15.0	14.1
6 Thailand	Phrapradaeng	29.3	73	1335	6.2	1.5	59	24	54	281	5.7	17.1	16.4
7 Vietnam	Hanoi	24.7	79	1556	5.8	0.8	15	18	49	182	6.1	5.4	16.8
8 Vietnam	Ho Chi Minh	28.3	74	1441	6.2	0.9	21	18	47	164	6.9	8.0	6.7
9 Vietnam	Mytho	27.0	81	1222	6.4	0.3	2	9	36	167	4.4	12.1	9.4
10 China	Chongqing	18.5	70	1162	4.5	1.3	99	45	52	783	9.1	24.2	31.2
11 China	Tie Shan Ping	18.5	90	1133	4.2	1.8	51	10	71	492	11.6	17.9	31.1
12 China	Hong Kong	22.9	78	2092	4.6	1.8	16	50	31	151	6.4	6.7	18.2
13 Malaysia	Kuala Lumpur	28.0	78	2776	4.3	3.8	12	47	42	139	8.1	9.6	21.6
14 Malaysia	Tanah Rata	18.1	91	2433	5.1	0.1	0	1	35	51	7.5	10.5	12.9
15 S. Africa	Johannesburg	17.2	78	417	4.8	2.1	18	28	51	105	2.0	4.6	31.4
16 Zambia	Kitwe	22.6	58	1083	4.7	0.9	92	11	72	464	27.1	12.7	35.0
17 Zambia	Magoye	22.2	62	826	7.0	0.5	0	2	53	27	2.0	5.1	8.3
18 Zimbabwe	Harare	18.9	63	798	6.6	0.7	16	15	65	193	3.5	4.1	8.3

The corrosion data after one year of exposure are summarized in Table 3.11. Regarding zinc one should note the slightly higher corrosion values for zinc at the Tie Shan Ping test site compared to Chongqing, which has a much higher SO$_2$ value, and also the very high values observed in Kitwe. For limestone the high values in Johannesburg, considering the pollution situation, deserves to be mentioned.

the values at polluted sites in Europe and illustrates the risk HNO_3 can play with increasing traffic in warm regions. The corrosion rate of steel is even lower in Tanah Rata than at the rural sites in the temperate region.

The final evaluation of results from the 4-years exposure from RAPIDC which will lead to dose-response functions is yet to be undertaken. In the meantime a comparison has been made with the observed 1-year values from Asia and Africa and predicted values from dose-response functions developed within other projects. In general the values predicted from the ICP Materials

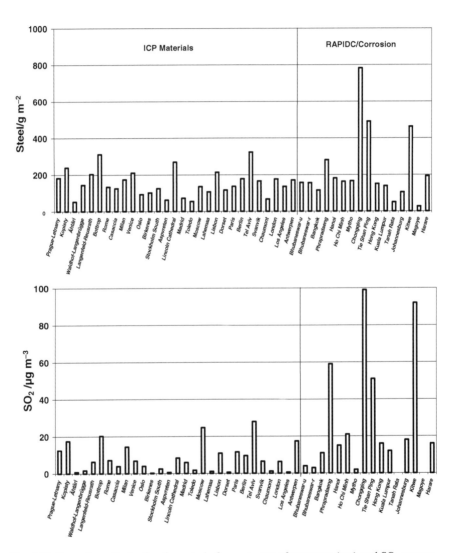

Fig. 3.23 Corrosion attack of carbon steel after one year of exposure (*top*) and SO_2 concentration (*bottom*) in the ICP Materials multi-pollutant programme and the CORNET networks

dose-response functions for the SO_2 dominating situation or ISO functions have a higher correlation to the RAPIDC data, which is not surprising considering the high SO_2 values in the CORNET network. This can be illustrated by a comparison of carbon steel corrosion and concentrations of SO_2 in the ICP Materials and RAPIDC networks, Fig. 3.23.

Preliminary conclusions in anticipation of results after 2 and 4 years of exposure, which will permit derivation of dose-response functions for subtropical and tropical regions, show that the corrosion values are higher than expected, except for zinc where the values are lower than expected using the best available dose-response functions from the temperate climate zone.

3.8 Concluding Remarks

An overall analysis based on results from different exposure programmes and on so far available data from the CORNET programme shows that high corrosion rates for metals and for calcareous stone materials, significantly higher than in Europe and Northern America, can be found in subtropical and tropical climates. This is due to a combination of several factors including higher SO_2 and HNO_3 and climatic effects. As many developing countries are located in warm regions with high relative humidity and high volume and frequency of precipitation, there is an obvious risk that corrosion due to deposition of acidifying pollutants will be more severe than in the temperate zone under similar emissions of pollutants. Similar to the temperate zone, dry deposition of SO_2, HNO_3 and particulate matter, "acid rain" and dry deposition of sea salt aerosols are the main parameters that influence the corrosion in tropical and subtropical regions. The relative importance of these three agents may, however, be different depending on the climatic conditions:

- The sensitivity of corrosion to dry deposition of SO_2 is similar in subtropical/tropical climates and temperate climate.
- The sensitivity of corrosion to wet deposition of H^+ (acid rain) is significantly higher in subtropical/tropical climates compared to temperate climate.
- The corrosion caused by deposition of sea salt chlorides increases substantially with temperature and can reach very high values in subtropical/tropical climates.

3.9 Case Study: Corrosion of Bronze Statues in the United States

3.9.1 Summary

More than a century of observations have linked atmospheric chemicals with increased copper and bronze corrosion. This section considers the corrosion of the complex forms of bronze sculpture, the applicability of dose-response functions for predicting corrosion of heritage bronzes, and the implications of both for designing bronze conservation treatment programs.

Replicates of a single-figure bronze statue, known as the Kitson *Hiker,* are widely distributed on sites across the United States, having been exposed to a range of industrial, urban, rural, desert and coastal environments for 3–7 decades. Kitson's *Hiker* statues are attractive corrosion monitors because of their fixed, complex geometry and reasonably consistent alloy chemistry, and are a useful, if serendipitous tool for evaluating the real-world applicability of the ICP Materials dose-response functions derived from standard bronze coupons. To place *Hiker* corrosion data in a 21st century cause and effect context, estimates of exposure histories dating back to the early 1920s were extrapolated from often sparsely available data. As a result of disconnects between advancing pollution measurement technologies, these exposure histories should be considered impressionistic.

Despite the quality of the aerometric data, the ICP Materials Multi-Pollutant dose-response function yielded annual corrosion rate estimates in very close agreement with corrosion measured on *Hiker*s in small cities and rural areas, but tended to significantly over-predict corrosion rates where SO_2 levels exceed $15 \ \mu g \ m^{-3}$. In contrast, the ICP Materials SO_2-dominant dose-response function yielded less accurate, but mostly reasonable estimates of life-time corrosion across the full range of *Hiker* exposures. Of ongoing interest for cultural property managers is the potential for surface moulding and profiling techniques to non-destructively documenting corrosion over time and tracking the effects of bronze conservation treatments.

3.9.2 Background: Corrosion of Copper and Bronze

Statuary bronze is a copper-tin alloy in a roughly 10:1 ratio; the typical bronze alloy used for American statuary is nominally 85% copper, 5% tin, 5% zinc, and 5% lead. Corrosion of bronze alloys is a more complex and slower process than the corrosion of copper. Bronze alloys used in outdoor sculpture are among the most corrosion resistant metals. Nonetheless, outdoor bronze sculpture left uncoated form dark or black corrosion films in sheltered, unwashed areas, while pitting and green streaking corrosion are found on skyward-facing, rain-washed surfaces.

The "greening" or *verdigris* patina formation on copper roofs is a well-known corrosion phenomenon (Graedel et al., 1987) and (Sherwood, et al., 1990). Copper ions interact with the environment to form a sequence of mineral corrosion products, resulting in a chemical composition reflective of the integrated, cumulative exposure. In the late 1920s, Vernon and Whitby (1929) observed that *verdigris* samples from rural copper roofs were composed of copper hydroxides, as compared with samples from urban roofs dominated by copper sulphates and samples from seaside roofs dominated by copper

Table 3.12 Copper corrosion at NAPAP test sites

NAPAP Test Site	Rain annual acidity pH	SO$_2$ annual average $\mu g\, m^{-3}$	Overall corrosion rate $\mu m\, yr^{-1}$	Sulphur related corrosion (%)	Acidity related corrosion (%)	All other corrosion factors (%)
Newcomb, NY	4.36	4.4	0.37 +/− 0.14	10	25	65
Washington, DC	4.21	27.2	0.83 +/− 0.19	38	25	37
Steubenville, OH	4.06	56.5	0.88 +/− 0.29	57	20	23

(Cramer et al., 1990)

chlorides. In the North-eastern United States, when ambient urban SO$_2$ levels of more than 100 $\mu g\, m^{-3}$ were common from 1900–1932, the "greening" rate of copper roofs was estimated to be 10–14 years. Similarly, increased rates of "greening" have been documented in Copenhagen. In the 1930s, Danish roofs became green in 20–30 years, compared to greening in 10 years in the 1950s and 8 years in the 1960s. The acidity of Copenhagen's rain observed in 1944–5 was pH 6 to 8, compared with pH 4 to 5 in 1965.

Late 20th century observations in the United States attribute 20–25% of copper corrosion to rain acidity (Table 3.12). Copper corrosion rates at a rural mountain test site are less than half of those found at the urban sites in Washington, DC and the heavily industrial Steubenville, OH, where sulphur oxides plays a dominant role in enhanced urban copper corrosion.

The sulphur content of corrosion films retained on sheltered, unwashed surfaces helps to document pollution's role in bronze corrosion. Lins and Power (1991) found that both wet and dry atmospheric deposition play important roles in bronze corrosion. Dry deposition of sulphur and nitrogen supplies reactive cations and anions to the outer zone of corrosion films. With increasing acidity, the removal of copper sulphate corrosion products by rain is enhanced, thus increasing both the corrosion of bronze statuary surfaces as well as the staining of adjacent stone and masonry materials. Copper stain removal from stone is one of the more difficult tasks facing conservators, suggesting that this secondary effect of pollution may be non-trivial.

3.9.3 Hiker Study – In Situ Measurement of Long-Term Monumental Bronze Corrosion

Theo Alice Ruggles Kitson sculpted a series of military figures at the turn of the 19th century. The *Hiker* figure represents an infantry soldier from the Spanish American War (1898–1902), the United State's first overseas conflict, where U.S. soldiers and sailors fought in the Spanish colonies of Cuba, Puerto Rico,

Fig. 3.24 Distribution Of Kitson Hiker Statues (1921–1965). Dashed lines plot rain acidity in units of pH, drawn from 1995 National Acid Deposition Network data, courtesy of the U.S. Geological Survey and the Colorado State University

the Philippines, as well as the Boxer Rebellion in China. Fiftytwo replicates of a single-figure bronze statue known as Kitson *Hikers* were cast by the Gorham Bronze Company and placed in cities across the United States in the 1920–1950s (Fig. 3.24). The suite of 50 Kitson *Hiker* figures offer an unusual perspective on bronze corrosion as these statues document cumulative corrosion of complex shapes for exposures of 30–74 years. Kitson's first *Hiker* statue (1906), located in Minnesota is missing its gun tip, the standard site for corrosion moulding, and thus is not included in this study. We were unable to gain access to the *Hiker* statue in New Orleans that stands atop a 15m column. The *Hiker* study group included fifty (50) *Hiker* statues, of which 16 figures had received conservation treatment prior to the study. The core of the *Hiker* corrosion study are the remaining 34 *Hiker* untreated statues. Kitson's *Hikers* are attractive corrosion monitors because of their fixed geometry, reasonably consistent alloy chemistry, wide distribution, and long period of exposure.

The Gorham Bronze Company cast Theo Alice Ruggles Kitson's *Hiker* figure at least 50 times between 1921 and 1965. Each time Gorham used the same pattern to sand cast about 1,000 pounds of statuary bronze. Metallurgically, the replicates are very consistent – dense, fine-grained alloys of between 88% and 90% copper with few detectable casting flaws. Metal samples were drilled from the gun butt of seven *Hiker* statues for analysis of alloy composition using x-ray reflection analysis. The figures were cold-patinated medium statuary brown and varnished. One-third of the *Hiker* statues had been conserved prior to this corrosion study, and thus not suitable for quantitative corrosion analysis. However, profiles of treated surfaces are instructive of the morphological changes

associated with various cleaning and coating technologies. The 34 unconserved
Hikers were considered in the corrosion rate analyses.

Kitson *Hikers* are found in nearly every climate zone in the continental
United States, which covers Desert to Alpine regimes. *Hikers* dedicated before
World War II are mostly found in the North-eastern United States; later statues
are almost all found in Southern and Western cities. *Hikers* have seen pollution
levels rise and fall dramatically, peaking in the 1930s and 1950s. All the *Hikers*
pre-date the United States Clean Air Act of 1970, which caused air pollution
levels in all cities to drop further by at least 50% and in some cases by an
additional factor of 5–10. The change in air quality was most dramatic in the
largest industrial cities, such as Chicago, New York, Philadelphia, Cleveland,
Indianapolis, Baltimore, and Boston, where SO_2 levels in the 1930s and 1950s
regularly exceeded 400 $\mu g\,m^{-3}$ annually.

The areal extent of green corrosion products of the Kitson whereas *Hiker*
statues was mapped and the statue's hydrology (the movement of water over
sculptural surfaces) was found to be remarkably diagnostic. Among all *Hikers*,
there is a consistent general pattern of bronze corrosion for rain-washed and
for sheltered surfaces. This washed-unwashed dichotomy is illustrated in the
composite map of *verdigris* corrosion on 25 of the *Hiker* replicates (Fig. 3.25).
Verdigris corrosion is consistently found on skyward facing, washed areas
of the statue, it is less frequently found on ground-facing, rain-protected surfaces. As the statue ages, *verdigris* corrosion covers a greater fraction of the
surface, expanding into rain-protected areas.

Texture of the bronze surfaces was measured with dental moulding putty, a
dimensionally stable vinyl polysiloxane. The putty was used to make very
precise replicas of corroded bronze surfaces in the field. The variability
among 4 sets of profiles from the same moulded surface, measured as a standard
deviation, ranges from 5% to 38% with a mean value of 25%. In view of the
three-fold difference between smoothest and roughest surfaces, a measurement
tolerance of about $+/-$ 25% is quite adequate to allow meaningful comparisons between *Hikers*. In the laboratory, a profilometer was used to trace
the surface character, and to quantify both pit depth and surface roughness
(Fig. 3.26). Surface texture measurements are translated to metal loss by using a
density of 2.8 $g\,cm^{-3}$ (Stöckle and Krätschmer, 1999).

The general trend of pit depth with the age of the statue is shown in Fig. 3.27
where maximum pit depth is plotted against years of exposure. A similar trend
was found in the relationship between surface roughness and dedication year.
This similarity indicates a relationship between surface roughness and maximum pit depth but this has not been explored in detail. The most corroded
Hikers exhibit total pit depths of 0.330–0.389 mm. Annual corrosion rates of
unconserved *Hiker* statues in the Northeast and Midwest United States., a region
known as the "Rust Belt" with typical rainfall volumes of 1 m y^{-1}, averaged
3.6 $\mu m\,yr^{-1}$. In contrast, for *Hikers* in the desert West, exposed to less than 0.5 m
rain per year, corrosion averaged 1.8 $\mu m\,y^{-1}$. The strong influence of rain,
measured either as total volume or total number of rain days (rain volume

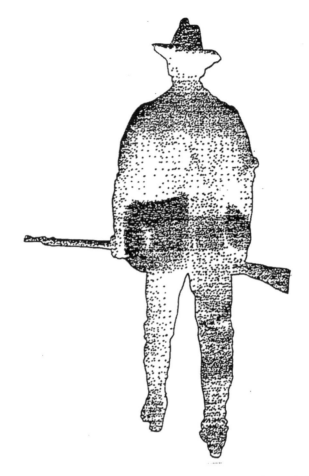

Fig. 3.25 Verdigris corrosion on 25 Kitson Hiker statues, courtesy of T. Meirding, University of Delaware, Geography Department

> 25 mm/day), on pit depth is illustrated in Fig. 3.28. The scatter of data in Figs. 3.27 and 3.28 reveals that neither age nor exposure to rain alone explains the degree of corrosion measured by maximum pit depth. Other exposure factors, such as air pollution and surface treatment (including vandalism and its repair) have a strong influence on the surface texture of exposed bronzes.

Exposure histories were estimated for all the *Hiker* locations, based on government data collected between 1920s and 1990s. Weather data are the most reliable, based on more than 50 years of data selected from 20,000 stations across the United States. In contrast, routine observation of SO_2 in the United States began in the late 1960s at 300 sites, and the first comprehensive reports of Air Quality trends date from the early 1970s (The National Air Monitoring Program: Air Quality Emissions and trends Annual Report 8/1973 EPA 450/

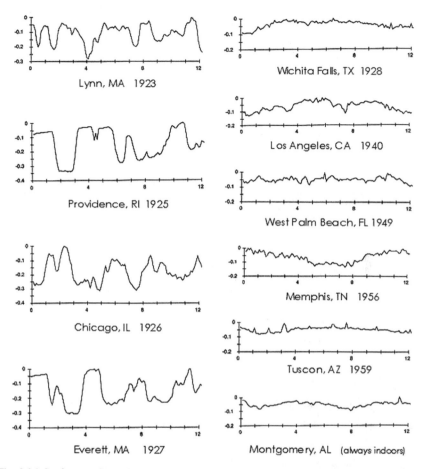

Fig. 3.26 Surface profiles of untreated Hiker statues 1923–1956, measurement taken at gun tip. The Montgomery, AL *Hiker, lower right*, always been indoors, provides a baseline surface texture. Courtesy of J. D. Meakin (retired), University of Delaware, Mechanical Engineering Department

1-73.001.á, EPA 40/1-73.001b, EPA 450/1-74.007). Since the mid-1980s, rain chemistry has been routinely monitored in 100 rural and remote locations; to date, rain chemistry has been measured only in a dozen U.S. cities. Particulate matter monitoring technology has evolved dramatically since the "smoke studies" of the early 1900s. Observations of air-borne particulate matter are more widely available than H^+ or SO_2, but these particulate data are particularly difficult to extrapolate beyond the measurement site and to compare with other measurement methods. Despite the uncertainty in estimating lifetime SO_2 exposures, the relationship between SO_2 and maximum pit depth is stronger than that for rain exposure.

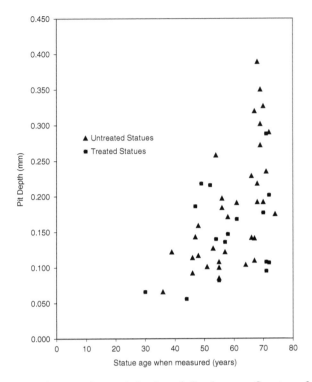

Fig. 3.27 Hiker corrosion – maximum pit depth vs. dedication year. Courtesy of J. D. Meakin (retired), University of Delaware, Mechanical Engineering Department

The 1920s *Hikers* located in rainy cities with heavy industry show corrosion rates nearly double the average for the era; i.e. Providence (1925, pit depth 0.327 mm), Chicago (1926, pit depth 0.350 mm), and Everett (1927, pit depth 0.389 mm). For this subset of *Hikers*, it appears that industrial density rather than urban population is more predictive of bronze corrosion. From 1920 to 1950, Everett's population was under 50,000, Providence's about 250,000, and Chicago more than 3,000,000. Chicago and Providence are two of the most industrialized *Hiker* cities; along with Baltimore and Los Angeles, these four *Hiker* cities were home to more than 1,000 manufacturers at their industrial peaks. Corrosion of *Hikers* in small industrial towns surrounding Boston, such as Everett (1927, pit depth 0.389 mm), Haverhill (1926, pit depth 0.302 mm), Lynn (1923, pit depth 0.290 mm), Wakefield (1926, pit depth 0.272 mm), and Waltham (1928, pit depth 0.320 mm) reflect exposure to the elevated SO_2 concentrations in the Greater Boston industrial air mass.

Hikers from the 1920s near Albany, NY, help illustrate the incremental role of industry within a small city. Pit depths are about half those observed in highly industrial areas described above. The Schenectady *Hiker* (1921, pit depth 0.175 mm) is located in a park in the suburban area west of the city,

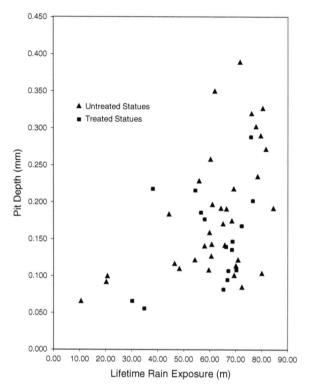

Fig. 3.28 Hiker corrosion – lifetime precipitation vs. maximum pit depth. Courtesy of J. D. Meakin (retired), University of Delaware, Mechanical Engineering Department.

while the Troy statue (1925, pit depth 0.192 mm) stands in a park adjacent to the industrial zone. Until World War II, industry prospered in the Albany area; Troy was a major 19th century-early 20th century steel producer. A third Albany statue, in the Cohoes industrial zone near Troy (1923, pit depth 0.202 mm) was aggressively cleaned to a "penny bright" surface in the early 1980s, and thus the surface loss may be as much the result of cleaning as of chemically aggressive atmospheric exposure. Cleaning removes corrosion products, so the measured surface textures may indirectly reflect the corrosion, in some undefined manner.

The younger *Hikers* dedicated in the 1940s and 1950s are located in a broader range of exposures than the "Rust Belt" concentration of the early *Hikers*. Most of the *Hikers* in coastal and desert exposures are in the younger set. Similar to the older statues, *Hikers* in manufacturing cities in the East Coast megalopolis evidence higher corrosion rates compared with statues in areas with less population and industry – compare Meriden, Connecticut (1941, pit depth 0.258 mm) and Cambridge, Massachusetts (1947, pit depth 0.159 mm) with

Knoxville (1940, pit depth 0.082 mm) and Kansas City (1947, pit depth 0.117 mm). Note that these four statues are thought to have received wax or other treatment at some point prior to surface profiling.

3.9.4 Predictive Power of Bronze Dose-Response Functions

Bronze corrosion rates calculated for multiple U.S. locations using ICP Materials dose-response functions result in highly plausible rates of statuary corrosion (Tables 3.13 and 3.14).

To illustrate the scale of bronze corrosion, loss calculations were made for the urban and rural sites in the Northeastern and Midwest United States using the ICP Materials SO_2-dominant dose-response function (Table 3.14). The damage is expressed as bronze thickness loss, in $g\,yr^{-1}$, also specified as loss of metal thickness in $\mu m\,yr^{-1}$. A bronze mass loss of $3\,g\,m^{-2}$ from cast bronze is equivalent to a

Table 3.13 Cast Bronze corrosion rates in the US estimated w/ICP materials SO_2-dominant dose-response function (Tidblad et al., 1999). Sample calculations using environmental data from 1997–2001

Location	Rain mm	Cl^- mg/l	H^+ mg/l	$SO_2\ \mu g\,m^{-3}$	Bronze loss μm/yr	Years to 0.1 mm Bronze loss
Data source	NADP, NYDEC	NADP, NYDEC	NADP, NYDEC	CASTNet NYDEC, EPA		
Dyken Pond, rural New York	963.6	0.127	0.035	4.19	1.86	54 yrs
Greater Albany, New York	830.2	0.164	0.044	18.3	3.25	31 yrs
West Point, New York	712.7	0.284	0.040	11.7	2.67	37 yrs
Buffalo, New York	936	0.200	0.037	16.7	3.13	32 yrs
Chicago, Illinois	989.6	0.203	0.021	24.7	3.50	29 yrs
Bondville, rural Illinois	854.4	0.164	0.024	6.3	2.02	50 yrs
Dunes, rural Indiana	1,127	0.164	0.021	6.3	2.06	48 yrs
Purdue Center, rural Indiana	824.9	0.160	0.028	10.8	2.53	40 yrs
St. Louis. Missouri	1,225	0.186	0.026	20.2	3.34	30yrs

Table 3.14 Cast Bronze corrosion rates – comparison of measured pit depth in Hiker statues with corrosion rates predicted by ICP Materials SO$_2$-dominant and multi-pollutant dose-response functions

City	State	Statue age years	Pit depth total mm	Pitting ann rate mm/yr	Annual precip mm	SO$_2$ 1970s µg m^{-3}	Multi DRF annual mm/yr	% Predicted pit depth multipol	ICP DRF cumm mm	% Predicted pit depth ICP DRF
Sacramento	CA	46	0.092	0.0020	439	3	0.001	46%	0.047	51%
Los Angeles	CA	55	0.1	0.0018	375	6	0.001	79%	0.082	82%
Portland	ME	71	0.235	0.0033	1105	6	0.003	83%	0.126	54%
Wichita Falls	TX	67	0.11	0.0016	721	6	0.001	75%	0.074	68%
Oshkosh Green Bay	WI	56	0.184	0.0033	791	7	0.003	80%	0.111	60%
West Palm Beach	FL	46	0.114	0.0025	1524	7	0.002	78%	0.059	52%
Malden	MA	57	0.122	0.0021	1243	8	0.003	127%	0.107	88%
Medford	MA	68	0.192	0.0028	1243	8	0.003	96%	0.126	65%
Wakefield	MA	69	0.272	0.0039	1185	8	0.003	68%	0.127	47%
Waltham	MA	67	0.32	0.0048	1137	8	0.003	56%	0.124	39%
Tucson	AZ	36	0.066	0.0018	295	9	0.001	43%	0.013	19%
Haverhill/Lawrence	MA	69	0.302	0.0044	1129	12	0.004	84%	0.156	52%
Manchester/Concord	NH	66	0.142	0.0022	997	12	0.004	192%	0.149	105%
Savannah	GA	64	0.104	0.0016	1250	12	0.002	149%	0.114	110%
Birmingham	AL	51	0.101	0.0020	1361	14	0.003	149%	0.112	111%
Cambridge	MA	48	0.159	0.0033	1247	15	0.004	127%	0.119	75%
Chelsea	MA	61	0.191	0.0031	1090	15	0.004	124%	0.159	83%
Everett	MA	68	0.389	0.0057	1054	15	0.004	68%	0.177	45%
Lynn	MA	72	0.29	0.0040	1107	15	0.004	105%	0.170	58%
Grand Rapids	MI	67	0.141	0.0021	865	16	0.005	258%	0.207	147%
Shamokin/WilkesBarre	PA	56	0.197	0.0035	1088	20	0.005	137%	0.158	80%

Table 3.14 (continued)

City	State	Statue age years	Pit depth total mm	Pitting ann rate mm/yr	Annual precip mm	SO_2 1970s $\mu g\,m^{-3}$	Multi DRF annual mm/yr	% Predicted pit depth multipol	ICP DRF cumm mm	% Predicted pit depth ICP DRF
Kansas City	MO	48	0.117	0.0024	968	22	0.005	204%	0.161	137%
Elmira	NY	66	0.229	0.0035	846	25	0.008	237%	0.234	102%
Memphis	TN	39	0.122	0.0031	1392	25	0.005	150%	0.109	89%
Morristown	NJ	47	0.143	0.0030	1291	25	0.005	180%	0.152	106%
New Bedford	MA	45	0.186	0.0041	1186	26	0.007	167%	0.150	81%
Portsmouth/Norfolk	VA	53	0.127	0.0024	1143	26	0.005	228%	0.161	127%
Providence	RI	70	0.327	0.0047	1151	26	0.007	139%	0.250	76%
Lebanon	PA	55	0.108	0.0020	1083	31	0.007	342%	0.189	175%
Pottsville/Sharon PA	PA	68	0.218	0.0032	1019	31	0.007	208%	0.230	105%
Meriden	CT	54	0.258	0.0048	1118	40	0.009	198%	0.238	92%
Troy	NY	70	0.192	0.0027	919	52	0.014	521%	0.320	167%
Allentown	PA	58	0.171	0.0029	1123	57	0.013	451%	0.304	178%
Chicago	IL	69	0.35	0.0051	899	65	0.017	327%	0.386	110%
Schenectady	NY	74	0.175	0.0024	926	68	0.018	779%	0.380	217%

corrosion layer 1.07 μm deep. By comparison, a typical piece of paper is 100 μm thick. Bronze thickness loss is a measure of surface recession that assumes a uniformly deep corrosion across the exposed surface, rather than pitting, where corrosion proceeds intensely in specific areas. For illustration purposes, the time it would take for a fully exposed bronze surface to recede the thickness of a sheet of paper (0.1 mm) was also computed for these locations. Setting a tolerable corrosion threshold of 0.1 mm is somewhat arbitrary and for these exposure regimes, the probable error on the estimates can be in excess of the computed differences. Error in these estimates is unknown, but may be as high as $+/-0.5$ μm, suggesting that small site-to-site differences should be evaluated with caution.

Bronze corrosion rates of $1.8-3.5$ μm yr^{-1} were estimated for urban and rural areas in the U.S. Midwest and New York State. The Midwest sites offer a range of SO_2 exposures similar to New York, but with generally lower rain acidity and higher rain volume (Gatz, 1991). The corrosion rates estimated with the ICP Materials SO_2-dominant dose-response function are in good agreement with the measured long-term corrosion rates of Kitson *Hikers*.

Further, the two ICP Materials dose-response function formulations were evaluated for the 34 unconserved *Hiker* statue exposures. To place *Hiker* corrosion data in a 21st century cause and effect context, estimates of exposure histories, dating back to the early 1920s were extrapolated from the often sparsely available data. As a result of disconnects between advancing pollution measurement technologies, these exposure histories should be considered impressionistic. Nonetheless, the multi-pollutant dose-response function yielded an estimated annual corrosion rate in very close agreement with corrosion measured on *Hiker's* in small cities and rural areas, while tending to significantly over-predict corrosion rates where SO_2 levels exceed 15 μg m^{-3} (Fig. 3.29). In the low-SO_2

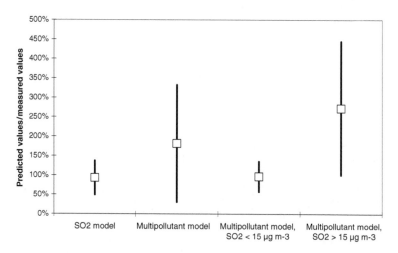

Fig. 3.29 Hiker pit depth measurements relative to. corrosion predicted by dose-response functions

cases, the multi-pollutant dose-response function predicted corrosion within 96% of the measured annual corrosion rate, with a standard deviation of 39%. The SO_2-dominant dose-response function formulation was evaluated with SO_2 values ten times greater than urban values reported in the early 1970s. In Europe, the percent reduction in SO_2 pollution over a fixed time was about the same when looking at both low polluted sites and high polluted sites, with an overall decrease of about 50% in the most recent ten-year period. For a thirty-year period (1970–1940) this would correspond to a factor of eight, which approaches the factor of ten differences used for the *Hiker* calculation. This formulation of the cast bronze dose-response function yielded quite reasonable estimates of life-time corrosion across the full range of *Hiker* exposures, although less accurate than those calculated with the multi-pollutant dose-response function for low SO_2 exposures. The mean predicted corrosion was 93% (std. dev 44%) of the measured corrosion; the predictions ranged from 19% to 217% of the measured corrosion. In other words, the SO_2-dominant dose-response function can under predict corrosion by a factor of five and over predict by a factor of two.

Acknowledgments The data and knowledge presented in this chapter are the result of an international co-operation between several organizations through three main efforts, the International Co-operative Programme on effects on materials including historic and cultural monuments (ICP Materials) under UNECE and the Convention on Long-range Transboundary Air Pollution, the EU 5FP MULTI-ASSESS and the EU 6FP CULT-STRAT. In particular, the following individuals and organizations are gratefully acknowledged:

— Dagmar Knotkova and Katerina Kreislova of SVUOM, Prague, Czech Republic for providing the ICP Materials sub centre for carbon steel, weathering steel, zinc and aluminium
— Rolf Snethlage of the Bavarian State Department of Historical Monuments, Munich, Germany for providing the ICP Materials sub-centre for copper and cast bronze including pre-treated bronzes.
— Tim Yates of the Building Research Establishment (BRE), Garston, Watford, United Kingdom for providing the ICP Materials sub-centre for limestone and sandstone.
— Jan Henriksen and Terje Grøntoft of the Norwegian Institute for Air Research (NILU), Lilleström, Norway for providing the ICP Materials sub-centre for coil coated galvanised steel with alkyd melamine, steel panel with alkyd, wood panel with alkyd paint and wood panel with primer and acrylate and the environmental sub-centre.
— Manfred Schreiner and Michael Melcher of the Institute of Chemistry, Academy of Fine Arts, Vienna, Austria for providing the ICP Materials sub-centre for glass materials representative of medieval stained glass windows including potash-lime-silica glass M1 (sensitive) and potash-lime-silica glass M3.
— Markus Faller and Daniel Reiss of EMPA, Corrosion/Surface Protection, Dübendorf, Switzerland, for providing the ICP Materials sub-centre for zinc.
— Stephan Fitz, Umweltbundesamt, Germany, for valuable discussions

The Swedish International development cooperation agency (SIDA) is acknowledged for financial support of the RAPIDC project and the organizations presented in Table 3.10 are gratefully acknowledged for their participation and performing all exposure in Asian and African countries.

References

Brimblecombe P, 1987. The big smoke – A History of Air Pollution in London since Medieval Times. London: Methuen.

Geike F R S, 1880. Rock weathering as illustrated in Edinburgh churchyards. Proceedings of the Royal Society of Edinburgh, 10: 518–532.

Knotkova D, 1993. "Atmospheric Corrosivity Classification. Results from the International Testing Program ISOCORRAG," Corrosion Control for Low-Cost Reliability, 12th International Corrosion Congress, Vol. 2, Progress Industries Plant Operations, NACE International, Houston, Texas, pp. 561–568.

Magnus G, 1864. The influence of bronze composition on the formation of an attractive green patina. Dingler Polytechnisches Journal 172: 371–376.

Mikhailov A A, Tidblad J and Kucera V, "The Classification System of ISO 9223 Standard and the Dose-Response Functions Assessing the corrosivity of Outdoor Atmospheres" Protection of Metals. 2004. Vol. 40. N 6. pp. 541–550 (In Russian variant pp. 601–610).

Morcillo M, Almeida E M, Rosales B M et al., Eds., "Functiones de Dano (Dosis/ Respuesta) de la Corrosion Atmospherica en Iberoamerica, "Corrosion y Proteccion de Metales en las Atmospherias de Iberoamerica, Programma CYTED, Madrid Spain, 1998, pp.629–660.

Zallmanzig J. 1985. Investigation on the rates of immision and effects in selected places of Europe for the quantitative examination of the influence of air pollution on the destruction of ashlar. Report of the NATO/CCSM Pilot Study on Conservation and restoration of Monuments, Number 158 Umwelbundesamt, Berlin.

Svensson, J-E and Johansson L-G, 1993, A laboratory study of the effect of ozone, nitrogen dioxide and sulphur dioxide on the atmospheric corrosion of zinc, J. Electrochem. Soc., Vol 140, No 8, pp. 2210–2216.

Schikorr G and Schikorr I, 1943. Über die Witterungsbeständigkeits des Zinks, Z. Mettallk-unde Vol 35, No 9, pp 175–181.

Brown, B. F., H. C. Burnett, W. T. Chase, M. Goodway, J. Kruger and M. Pourbaix. 1977. Corrosion and Metal Artifacts – A Dialogue between Conservators and Archaeologists and Corrosion Scientists. Washington, DC: National Bureau of Standards.

Cramer, S. D. and L. G. McDonald. 1990. Atmospheric factors affecting the corrosion of zinc, galvanized steel, and copper. In: *ASTM STP 1000 Corrosion Testing and Evaluation. Committee G1 Symposium*, eds. S. Dean and R. Baboian, Philadelphia, PA: ASTM.

Drayman-Weisser, ed. 1992. Dialogue/89- The Conservation of Bronze Sculpture in the Outdoor Environment: A dialogue among Conservators, Curators, Environmental Scientists, and Corrosion Engineers. Houston: NACE International.

Fitz S Ed., 1999. Quantification of Effects of Air Pollutants on Materials. Berlin: Umwelt-bundesamt (Federal Environmental Agency).

Gatz, Donald F. 1991. Urban Precipitation Chemistry: A Review and Synthesis. *Atmospheric Environment* Vol. 25B, no. no. 1: 1–15.

Graedel, T. E., Nassau, K., and Franey, J. P., 1987. Copper Patinas Formed in the Atmosphere - I. Introduction. *Corrosion Science*, 27(7): 639–657.

Lins, A. and Power, T. The Corrosion of Bronze Monuments in Polluted Urban Sites: A Report on the Stability of Copper Mineral Species at Different pH Levels. Scott, D. A., Podany, J., and Considine, B. B. Ancient and Historic Metal Conservation and Scientific Research. pp. 119–151. 1991. Marina del Ray, Getty Conservation Institute.

Lipfert, F. W. 1991. Historic Urban SO_2 Levels. *APT Bulletin* XXIII, no. 4: 72. Notes adapted from F. W. Lipfert, 1987, Estimates of historic urban air quality trends and precipitation acidity in selected U.S. cities (1880–90), Brookhaven National Laboratory Report 39845.

Sherwood, S. I., D. F. Gatz, Jr. R. P. Hosker, C. I. Davidson, D. A. Dolske, B. B. Hicks, D. Langmuir, R. Linzey, F. W. Lipfert, E. S. McGee, V. G. Mossotti, R. L. Schmiermund, and E. C. Spiker. 1990a. *Processes of Deposition to Structures.* Acidic Deposition: State of Science and Technology, ed. P. Irving, Vol. III, Report 20. Washington, D.C.: National Acidic Precipitation Assessment Program. (*SOS/T* 20).

Stöckle, B., and Krätschmer A. 1999. Quantification of Effects of Air Pollutants on Copper and Bronze. In: *Quantification of Effects of Air Pollutants on Materials.* ed. S. Fitz. Berlin: Umweltbundesamt (Federal Environmental Agency).

Vernon WHJ and Whitby L, 1929. Open air corrosion of copper, a chemical study of the surface patina. J. Institute of Metals, Vol. 42:181.

Weil, P.D., Naude, V.N. Patina, a historical perspective artistic intent and subsequent effects of time, nature and man. pp. 21–27. 1985. Philadelphia, Pennsylvania Academy of the Fine Arts.

Sources of Additional Information

General mechanisms and metallic materials
Leygraf C. and Graedel T. E. Atmospheric Corrosion, Electrochemical Society Series, ISBN 0-471-37219-6, John Wiley & Sons, Inc., 2000.

Metallic and non-metallic materials
Brimblecombe P., The effects of Air Pollution on the Built Environment, ISBN 1-86094-291-1, Imperial College Press, London, 2003.

Effect of chlorides
Dean S. W., Delgadillo G. H.-D and Bushman J. B., Marine Corrosion in Tropical Environments, ASTM STP 1399, ISBN 0-8031-2873-8, American Society for Testing and Materials, West Conshohocken, PA, 2000.

Effect of HNO_3
Samie F., HNO_3-induced Atmospheric Corrosion of Copper, Zinc and Carbon Steel, Thesis, ISBN 91-7178-483-7, Royal Institute of Technology, Stockholm, Sweden

Trends
Tidblad, J., Kucera, V., Mikhailov, A. A., Henriksen, J., Kreislova, K., Yates, T., and Singer, B., "Field Exposure Results on Trends in Atmospheric Corrosion and Pollution", Outdoor and Indoor Atmospheric Corrosion, ASTM STP 1421, H. E. Townsend, Ed., American Society for Testing and Materials, West Conshohocken, PA, 2002.

Tidblad J, Kucera V, Henriksen J, Kaunisto T, "Mapping and Trends of Acid Deposition Effects on Materials in Scandinavia", 13th Scand. Corros. Congr., (NKM13), Reykjavik, Iceland,. April 18–20, 2004.

Kucera, V., Tidblad J. and Yates T., "Trends of pollution and deterioration of heritage materials", Proc. 10th Int. Congr. Deter. Conserv. Stone, Stockholm June 27–July 2, 2004, Vol. 1, pp. 15–26.

An excellent overview of copper and bronze corrosion chemistry vis à vis pollution is found in:
Graedel, T.E., 1987. Copper Patinas Formed in the Atmosphere – II. A qualitative assessment of mechanisms. Corrosion Science, 27(7): 721–740.

Graedel, T.E., Nassau, K., and Franey, J.P., 1987. Copper Patinas Formed in the Atmosphere – I. Introduction . Corrosion Science, 27(7): 639–657

On CORNET and ICP Materials programmes
Tidblad, J., Mikhailov, A.,& Kucera, V. (2000). Acid deposition effects on materials in subtropical and tropical climates. Data compilation and temperate climate comparison. SCI Report 2000:8E, Swedish Corrosion Institute, Stockholm, Sweden.

Kucera, V., Tidblad, J. (2005). Comparison of environmental parameters and their effects on atmospheric corrosion in Europe and in South Asia and Africa. *Proc. 16th Int. Corrosion Congress,* Beijing.

Tidblad, J., Kucera, V., Samie, F. et al., (2007). Exposure Programme on Atmospheric Corrosion Effects of Acidifying Pollutants in Tropical and Subtropical Climates. *Water, Air and Soil Pollution: Focus* 7: 241–247.

On other exposure programmes

Callaghan, B. G. (1991). Atmospheric corrosion testing in southern Africa: results of a twenty- year national exposure programme. Division of Material Science and Technology, GAcsir 450H6025*9101, Scientia Publishers, CSIR, pp. 75.

Knotkova, D. (1993). Atmospheric corrosivity classification. Results of the international testing programme ISO CORRAG. In: *12th International Corrosion Congress,* vol. 2 (pp. 561–568). Houston, Texas: Progress in Industries Plant Operations, NACE International.

Morcillo, M., Almeida, E. M., Rosales, B. M, et al. (Eds.) (1998). Functiones de Dano (Dosis/ Respuesta) de la Corrosion Atmospherica en Iberoamerica, Corrosion y Proteccion de Metales en las Atmospheras de Iberoamerica, Programma CYTED, Madrid, Spain, pp. 629–660.

Cole, I. S. (2000). Mechanisms of atmospheric corrosion in tropical environments. ASTM STP 1399. In S. W. Dean, G. Hernandez-Duque Delgadillo & J. B. Bushman (Eds), *American Society of Testing and Materials.* West Conshohocken, PA.

Maeda, Y., Moriocka, J., et al., (2001). Materials damage caused by acidic air pollution in East Asia. *Water, Air and Soil Pollution,* 130, 141–150.

Chapter 4
Soiling

John Watt, Ron Hamilton, Roger-Alexandre Lefèvre, and Anda Ionescu

4.1 Overview

Soiling is another simple and familiar concept – we all know (or remember) the blackness of buildings associated with industrial urban cities or the streaked effect on limestone buildings. This chapter looks at the actual mechanisms involved and relates them to atmospheric pollution.

Soiling is a visual effect – the darkening of exposed surfaces due to air pollution – and may be said to come into the category of nuisance, and the extent of the nuisance caused by the soiling depends on when the response of the observer identifies the effect as unacceptable. The studies that underpin this will be introduced.

The scientific study of the complex processes resulting from particles of different types depositing on surfaces requires that a limitation is put on what is considered soiling. This is outlined in the introduction to the chapter, which also considers the resulting variability of the soiling pattern.

A number of studies have tried to examine the levels and trends in soiling by measurement of loss of reflectance of white surfaces or loss of transmission through glass. These, and the models derived for soiling rate are presented. A timely reminder that particulate matter is a complex mixture and that soiling has some very site specific influences on occasion is given in a brief outline of some scanning electron microscopy results from a study that looked at particles depositing on stable surrogate surfaces of the type used in the soiling rate studies.

The chapter ends with a presentation of some preliminary dose-response functions for soiling and discusses their use in policy making.

J. Watt (✉)
Centre for Decision Analysis and Risk Management, School of Health and Social
Sciences, Middlesex University, The Burroughs, London, NW4 4BT, UK
e-mail: j.watt@mdx.ac.uk

J. Watt et al. (eds.), *The Effects of Air Pollution on Cultural Heritage*,
DOI 10.1007/978-0-387-84893-8_4, © Springer Science+Business Media, LLC 2009

4.2 Introduction

Soiling is a visual effect resulting from the darkening of exposed surfaces following the deposition and accumulation of atmospheric particles. Deposition, removal and accumulation processes are numerous and complex, depending on the physical and chemical properties of the particles, the nature of the surface, the local meteorology and the pathways followed by rainwater after it hits the building surface. As a result of these complex interactions, there can be substantial variations in the level of soiling observed on building surfaces. These variations can be seen on the buildings shown in Figure 4.1, historic buildings in London, Belfast and Paris. Soiling is not uniform; there are variations across both horizontal and vertical levels. There is an overall trend of reduced soiling at higher levels on the building but this is not a smooth and continuous effect because of the number of variables involved.

Soiling is one of the effects of air pollution which comes into the category of nuisance and the extent of the nuisance caused by the soiling depends on the response of the observer (Grossi and Brimblecombe, 2004, Brimblecombe & Grossi, 2005). These studies used on-site questionnaires and found a clear relationship between opinions about the dirtiness of a building and views that it should be cleaned. The relationship was sufficiently strong for the authors to consider the establishment of aesthetic thresholds that could define levels at which the appearance of the building becomes publicly unacceptable. However, increased frequency of cleaning, washing, or repainting of soiled surfaces becomes a considerable economic cost and can reduce the useful life of the soiled material (Newby et al., 1991).

The distinction between soiling and some other forms of damage is, to an extent, artificial. The area where soiling and damage overlap most clearly is when black crusts are found on a building surface. Different authors use different terminology to describe these effects; in this book, *soiling* is used to describe the accumulation of particulate matter on a non-reactive surface and this accumulation can be removed by washing; *crusts* are taken to mean a compact surface layer with different chemical composition from the original material, formed by chemical weathering. Gypsum crusts are the most common type of growth found on building surfaces. Gypsum is calcium sulphate dihydrate, with the chemical formula $CaSO_4 \cdot 2H_2O$. Gypsum crusts are formed on calcareous stones following SO_2 deposition to the surface in the presence of moisture, followed by the dissolution of calcite and the precipitation of gypsum. Gypsum crusts have also been observed on calcite bound sandstones. The black colour of gypsum crusts is the result of the accumulation of particulate matter within the crust (Del Monte et al., 1981; Saiz Jimenez, 1993).

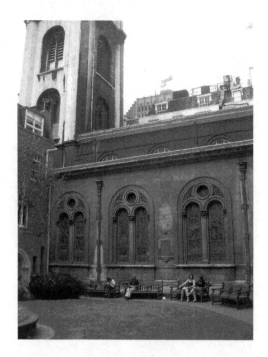

Fig. 4.1 Soiling on buildings in London, Belfast and Paris

Fig. 4.1 (continued)

4.3 Levels and Trends

This section reports observed levels of soiling reported in the recent literature, at sites which have more than a local significance. Most studies have investigated the variation of soiling (measured as a loss in reflectance) as a function of time at one or more spatially separated sites. These studies provide information on temporal trends. A small number of studies have concentrated on one building and investigated the variation of soiling with altitude, the so-called vertical trends.

4.3.1 Temporal Trends and Variations

Early studies of soiling were conducted at various locations in the United States of America. They were largely descriptive, reporting the occurrence of black specks on the freshly painted surface of a building in an industrial area (Parker, 1955) and the effects of particles on the painted exterior surfaces of homes

(Spence and Haynie, 1972). These studies reported that a consequence of soiling was that repainting was required over a 2–4 year time period, but this was under conditions of particulate concentration rising to several hundred μg m^{-3}. Beloin and Haynie (1975) introduced the idea that soiling could be measured using the contrast between reflectance of a soiled surface and reflectance of a bare substrate. Materials were monitored over a two year period at five sites with significantly different characteristics and with TSP concentrations ranging between 60 and 250 μg m^{-3}. The results were fitted statistically to an equation with reflectance loss (soiling) directly proportional to the square root of the dose, where dose was defined as the product of the particulate matter concentration and time of exposure.

Further studies by Haynie and Lemmons (1990) showed how various environmental factors contribute to the rate of soiling of white painted surfaces. Based on the results of this study, the authors concluded that coarse mode particles initially contribute more to soiling of both horizontal and vertical surfaces than fine mode particles; but the coarse particles were more easily removed by rain; Because of this, long-term remedial actions were more likely to be taken because of the accumulation of fine rather than coarse particles. Creighton et al. (1990) found that horizontal surfaces soiled faster than did the vertical surfaces and that large particles were primarily responsible for the soiling of horizontal surfaces not exposed to rainfall.

In Europe, a series of studies investigating the soiling of materials under conditions of high concentrations of traffic-emitted pollution in the absence of rainfall were undertaken by Mansfield and Hamilton (1989), who exposed white wood tablets and ceramic tiles for 250 days in a road tunnel. The results are shown in Fig. 4.2. These authors also placed the same type of white wood plates on the top of a roof in the London urban area over a period of 110 days. The results of both exposure experiments allowed the rate of soiling to be related

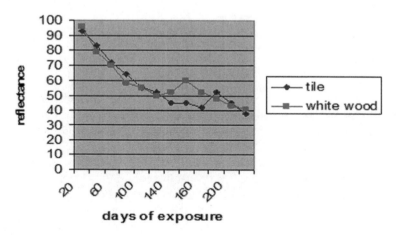

Fig. 4.2 Soiling (as % loss of reflectance) in a road tunnel

to atmospheric levels of black smoke and the prevailing meteorological conditions (Hamilton and Mansfield, 1992). In Portugal (Pio et al., 1998) sheltered surfaces showed a continuous decrease in reflectance, which followed a square root equation on exposure time. Their data predicted that a 30% decrease in reflectance would take between 5.5 and 8.8 years. They attributed 70% of the black carbon particles responsible for this soiling to vehicle emissions.

Temporal and spatial soiling patterns at the Cathedral of Learning, a 42-story National Historic Landmark on the University of Pittsburgh campus, were studied by a combination of archival photographs and computer modelling of rain impingement (Davidson et al., 2000). The photographs showed that the building had become soiled while still under construction. Surface cleaning by rain wash off was greatest at higher elevations and at the corner of buildings. It was concluded that rain washing of soiled areas on buildings occurs over a period of decades while the soiling process occurs more rapidly.

More recently, a trans-European project, MULTI-ASSESS (Swedish Corrosion Institute, KIMAB, 2006) measured soiling and corrosion patterns at a range of locations and gave detailed information on both temporal and spatial trends. The temporal trends in soiling over a 1-year period at locations in Krakow, for white plastic, are shown in Fig. 4.3.

The results confirm that there is significant variation in soiling from location to location and also that the trend is not continuous, reflecting variations in the location, local climate and pollution characteristics.

The soiling of modern glass has been the subject of a number of recent scientific studies, partly because of its widespread use in modern buildings and also because its smooth, non-porous, chemically inert surface makes it a good reference material (Lombardo et al., 2005). Samples of modern Si-Ca-Na float glass ($10 \times 10 \times 0.2$ mm) were exposed for up to 28 months sheltered and unsheltered from rain at five targeted exposure sites: four urban (London, Prague, Krakow, Athens) and a rural one (Monte Libretti, near Rome). Withdrawals were performed after 3, 6, 9, 12, 15, 18, 21, 24 and 28 months of exposure and four parameters were measured for each location/time. The total mass of deposited particles per surface unit (TP/S) was determined by the difference between the final and the initial mass of the exposed samples; the direct transmittance of light (T_{direct}) and the diffuse transmittance ($T_{diffuse}$) through samples were measured by spectrometry in the visible range. The haze was then determined: Haze % = 100 ($T_{diffuse}/T_{direct}$); the Total Carbon (Elemental Carbon + Organic Carbon) content of particles per surface unit (TC/S) was measured by a 2-step thermo-coulometric method; and the major soluble ions content of particles per surface unit (SI/S) was determined by rinsing samples with deionised water and analysing the solution by ion chromatography. Results for Krakow show that saturation is attained after 600 days (on average) and 700 days for Prague; but an increasing trend is still observable in the London samples after 750 days, those in Rome after 850 days and in Athens after 550 days.

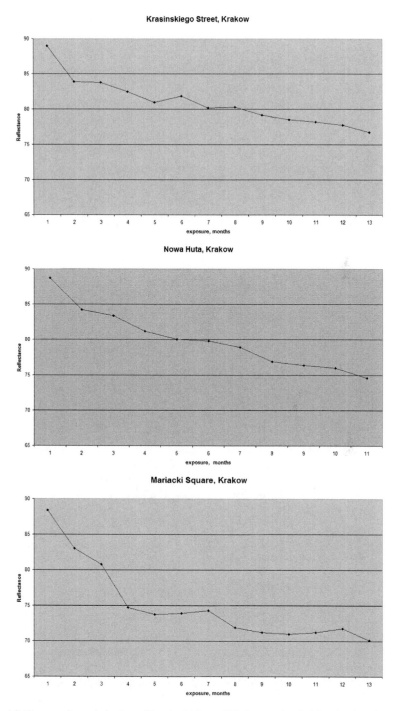

Fig. 4.3 Temporal trends in the soiling (as % loss of Reflectance) of white plastic at Krakow

4.3.2 Spatial Trends and Variations

The results presented in Fig. 4.3 give an indication of the significant spatial variations which are observed in soiling studies. This was shown on the international scale in field exposures performed within the MULTI-ASSESS project. Soiling measurements were recorded at a network of sites set up within the project. Measurements were taken at the beginning of the exposure programme and again after twelve months, to demonstrate any changes in reflectance and therefore provide a snapshot of 1-year soiling. The results are shown in Fig. 4.4.

It seems more appropriate to describe these results as spatial variations rather than spatial trends. Local conditions (local emissions and meteorological conditions) are likely to dominate and it is these factors rather than national trends which are responsible for the results shown in Fig. 4.4.

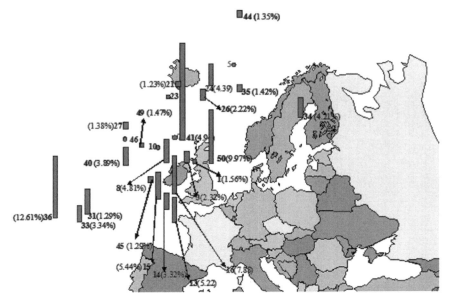

Fig. 4.4 1-year soiling recorded as a % loss of reflectance at a range of locations across Europe The sampling locations can be identified by the site numbers (MULTI-ASSESS (2007))

4.4 Mechanisms and Models

For a soiling model to be useful in policy terms, it should preferably have a physical basis rather than be solely statistical because this permits soiling rates to be described in terms of atmospheric pollution (especially particulate matter) concentrations and gives the model a more fundamental significance with application beyond the locations which provided the original data. The exponential model is the one which has been most commonly used in soiling studies applied to stone, wood and plastic because it meets this criterion. A model based on the Hill's equation has been found to be best for describing the soiling of glass.

The theoretical basis of the exponential model has been described by Mansfield and Hamilton (1989) and assumes that the level of reflectance is determined entirely by the percentage of surface that is covered by one or more particles. The mathematics of the model uses the following parameters:

A_0 total area of surface which receives depositing particulate
A uncovered area of surface which receives depositing particulates
r   radius of particles (geometric)
ρ density of particles
C concentration of particles in ambient air
v_d deposition velocity of particles
Y resuspension/removal rate

The rate at which mass deposits to uncovered surface $= A\,C\,v_d$
The rate at which mass is removed from uncovered surface $= A\,Y$

Combining these equations leads to

$$A = A_o \exp(-kt)$$

where

$$k = \frac{3(Cv_d - Y)}{4r\rho}$$

The loss in reflectance is related to the area covered as follows:

R_0 Reflectance from uncovered surface
R_p Reflectance from particles on covered surface

then

$$R = R_0\left(\frac{A}{A_0}\right) + R_p\left(\frac{A_0 - A}{A_0}\right)$$

$$= \left(\frac{R_0 - R_p}{A_0}\right)A + R_p$$

so

$$R = (R_0 - R_p)\exp(-kt) + R_p$$

the approximation $Rp = 0$ leads to the equation

$$R = R_0 \exp(-kt) \tag{4.1}$$

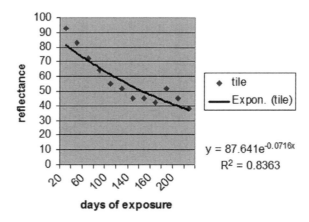

Fig. 4.5 Basic exponential model applied to soiling in a road tunnel

This is the relationship (often called the basic exponential relationship) which has been most frequently used in soiling studies.

Figure 4.5 shows this basic exponential equation applied to the data presented in Fig. 4.2. The fit is generally good ($R^2 = 0.8363$) but with a significant variation at the time 180–200 days. Variations of this nature are common in soiling data sets and reflect changes in meteorological conditions, cleaning or an unusual deposition event.

Haynie (1986) extended this approach by taking account of the particle size spectrum present in the ambient particulate matter and the dependence of deposition velocity on particle size. The soiling constant k was then evaluated by summing across the particle size range.

Lanting (1986) considered an alternative mathematical approach, modelling soiling as a consequence of the build-up of a thin film and light attenuation within this film. This is a classical situation in physics e.g. the absorption of X-rays by materials, and one which results in an exponential relationship.

A considerable debate surrounds the choice of the most appropriate measure of particulate matter to represent concentration C ($\mu g\ m^{-3}$) in the soiling models described above. Possibilities include Total Suspended Particulate matter (TSP), PM_{10}, $PM_{2.5}$ or Particulate Elemental Carbon (PEC). Light absorption by ambient particles is almost exclusively caused by PEC (Horvath, 1993) and PEC concentrations are due mainly to traffic (particularly diesel) emissions (Hamilton and Mansfield, 1992). A good argument can also be made that *particle deposition rate* is more appropriate than ambient concentration. However, the choice of parameter (TSP, PM_{10}, $PM_{2.5}$ or PEC) is generally governed by the information available from monitoring programmes.

A different approach to modelling was developed by the Vienna team (Aksu et al., 1996; Toprak et al., 1997; Pesava et al., 1999) who placed cubes in a wind tunnel to simulate ambient conditions. This arrangement allowed them to examine dry deposition of particles to building surfaces and soiling. The turbulent flow field around the surfaces was measured by laser Doppler anemometry. Two techniques were employed to simulate airborne particles; polydisperse soot particles with a mass mean diameter of 0.8 μm produced by atomizing a suspension of India ink and monodisperse fluorescent 0.6 μm latex spheres. The deposition velocities of the fluorescent spheres were measured with a fluorescence microscope. Measurements on the cubes exposed to soot particles recorded gave results for changes in reflectance. Spatial variations in deposition and soiling were observed, with the highest deposition always found on the edges of the cubes.

The selection of a model most appropriate for describing the soiling of glass was based on work carried out on data obtained from a 2-year exposure of silica-soda-lime (modern) glass in Paris (Lombardo et al., 2005). Glass soiling could not be analysed via the reflectance, but another optical property, appropriate to transparent materials, was measured: the haze, defined as the quotient between the direct and the diffuse components of the light transmittance. This study showed that the best fitting model was the Hill equation, corresponding to an S-shape curve.

The Hill equation, also known as the variable slope sigmoid, is the most frequently used model in pharmacology to analyse dose-response functions. It was originally derived by Hill to describe the uptake of oxygen by haemoglobin. Moreover, Hill equation found various applications in modelling adsorption phenomena in botany.

The general form of the Hill equation allows the variation of a measured parameter $Y(t)$ with time (t) to be expressed as:

$$Y(t) = B + K \cdot \left[1 + \left(M \cdot t^{-1} \right)^{H} \right]^{-1}$$

where:

B *(Bottom)*: initial level of response or, more generally, level of response in the absence of dose; in most of the cases is considered a priori null;

K *(Span)*: Top-Bottom, where Top is the value of the soiling parameter corresponding to the maximum curve asymptote (saturation), or level of response produced after infinite soiling;

M *(Half-life)*:time (dose) when the response is halfway between the Top and Bottom; it corresponds to the curve inflection;

H *(Hill slope)*:maximum slope of the dose-response curve at time M; it is used as a measure of the evolution rate.

Hill's equation is a particular case of the Verhulst model: it can be obtained if time is replaced in the Verhulst equation by its logarithm (Ionescu et al., 2006). The Verhulst equation models the population growth in a limited environment. It assumes that: resources restrict the growth, so the population reaches a stable size; the growth rate is proportional to its current size and to the unused environmental capacity. This model can be expressed mathematically as:

$$P(t) = \frac{K}{1 + \exp[-(\alpha \cdot t + \beta)]}$$

Where:

P : size of the population at time t,
α : intrinsic rate of natural growth,
K: carrying capacity of the environment,
β : location parameter.

Figure 4.6 shows the application of the model to results from Krakow, Poland. According to this model, soiling is a 2-kinetics phenomenon with variable soiling rate. During the first stage soiling rate increases to a maximum, then during the second phase, it decreases to zero on attaining saturation (after infinite soiling). The first stage corresponds to the capture of particles by the reactive sites present on the glass surface and its consequent progressive covering. During the second stage, the deposit mechanism changes because glass surface properties change.

Three other parameters (total mass of deposit, total carbon and soluble ions) revealed the same trend as the haze and led to the selection of the same best-fit model. The MULTI-ASSESS project offered the possibility to test and to validate this model using data obtained for other five sites involved in this project (Ionescu et al., 2006). The results showed the same trend as in Paris, but revealed more scatter around the S-shaped curve. The authors noted that this 6-site comparison

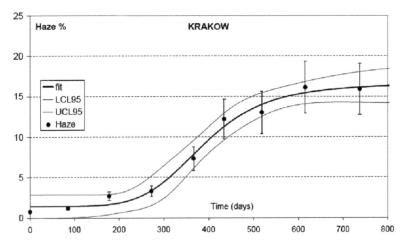

Fig. 4.6 Soiling of glass exposed in Krakow: measurements and Hill's model (Ionescu et al., 2006).

revealed that two model coefficients had similar values for all the 6 cases, while two others, reflecting the soiling amplitude, were related to the specific environment.

4.5 Variability in the Particle Assemblage

Although single dose-response functions, related to concentration of some measure of total particulate material or a size fraction of it, have been discussed to date, this represents a simplification of reality. Dose-response functions need to relate some measure of concentration that can easily be monitored if they are to have any practical use for policy making. This has been presented for pollutant gases discussed in Chapter 3 with respect to corrosion dose-response functions.

In the current case, however, things are rather more complicated. Particulate matter is a complex pollutant and is routinely monitored by local and national authorities based on size classification rooted in health studies (e.g. PM_{10}, $PM_{2.5}$). When considering the soiling potential of a particle, however, its other physical and chemical properties must be considered. It is thus recognized that the actual point at which soiling becomes perceptible depends on properties such as the colour and morphology of the particles and the optical properties of the surface of interest. In other words, particulate material is not a single pollutant but rather a complex mixture. Figure 4.7 shows, however, that there is, in general, a good relationship between deposition and loss of reflectance measured over a year, although there are some outliers.

As Fig. 4.8 shows, the bi-monthly samples show more scatter, as might be expected from a shorter sampling time, and include some sites that appear, at first sight, to conform less well to the modelled slope. The influence of site specific characteristics on the makeup of the particle assemblage

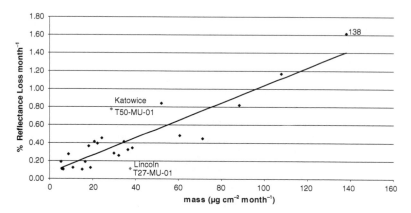

Fig. 4.7 The relationship between mass and reflectance loss at the MULTI-ASSESS broad field sites, annual samples

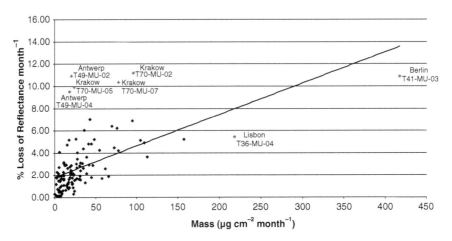

Fig. 4.8 The relationship between mass and reflectance loss at the MULTI-ASSESS broad field sites, bi-monthly samples

can be demonstrated by examining some of the particles from the sampling stations identified as outliers in Fig. 4.8 by microscopy and scanning electron microscopy.

Although many analytical methods can be adapted for the study of individual particles, the technique of particle characterisation by computer controlled scanning electron microscopy (CCSEM) has been widely used to characterise both emissions from pollution sources and examination of mixed samples from sites at which pollution is deposited (see Watt, 1998, for a review). The technique is applicable wherever there are particulate emissions that can be collected and analysed, with the important condition that such features contain information about their formation and source. This information can be used for source apportionment wherever the particulate material emitted by different processes differs between sources, or contains components that do so. Conventional analyses of the bulk sample represent average values for many such particles measured together and thus cannot provide information on the relative contribution of different emissions.

Any scanning electron microscope is an individual member of a family of different types of instrument, comprising several different potential microanalytical systems. In each case a beam of electrons is focused onto a specimen and a number of signals result, this type of study mostly uses three detected signals – secondary electrons, backscattered electrons and X-rays. Each may be used to gather images of a whole field of view in the instrument or used to gather data from individual particles if the scan of the beam is restricted to within its boundaries.

Secondary electrons are electrons that have been ejected from the specimen by the passage of a high energy electron from the beam of the instrument. They are collected in the secondary electron detector and the image they form is the one most often used for published micrographs. As such pictures show, these images contain a great deal of textural and morphological information. Secondary electron images suffer from one major disadvantage as far as automated particle analysis is concerned and that is, that the signal strength may be greatly affected by particle thickness, edge effects or charging. This means that different particles of the same composition may have very different brightness's in the image. Backscattered electrons (BSE) are high energy electrons from the incident beam of the SEM which have been diverted by a series of collision events in the target (sample). In contrast to the secondary electron images, the output signal strength is proportional to the mean atomic number (the Z number) and so images derived from this signal may readily be subdivided based on useful threshold values to isolate features of interest. X-rays are generated as a result of the release of energy which accompanies the movement of an electron from an outer shell to replace a dislodged secondary electron. The amount of each element in the target is given by the output of characteristic X-rays, measured by their wavelength or energy (often known as electron microprobe analysis EPMA). For the examples described here, data were gathered using energy dispersive X-ray spectroscopy (EDS).

Two types of samplers have been developed for use with computer controlled scanning electron microscopy, cylindrical and flat. The former sample deposition irrespective of direction and the latter mimic flat building surfaces. Teflon filters, as used by similar passive samplers optimised for bulk chemical analysis, are not suitable for microscopy as they do not retain particles at the surface. The parallel samplers, therefore, were constructed using Millipore Isopore Filters (HTTP 0037), which retain particles on the surface and which show excellent stability under the SEM beam. Both types of samplers act as inert passive samplers and so are directly equivalent to the samplers used for measurement of soiling.

Figure 4.9 is a digital image of the cylindrical sampler from the Berlin site, taken in a light microscope. In use the filter was wrapped around a cylinder, so that particles carried from all directions would deposit on it. The stripe of dark particles shows that one direction predominated and this site is located very close to a major road. The winter traffic uses "winter tyres" that are made of a soft

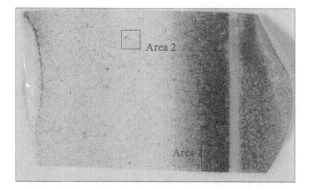

Fig. 4.9 Sample Berlin T 41-MU-03 (21/01/03-25/03/03). This winter sample shows that pollution was extremely directional

Fig. 4.10 Backscattered electron image from the scanning electron microscope of area 1 from the Berlin sample

compound that isn't affected by the cold but which is known to abrade more easily than many other types. This produces a winter peak of black rubber particles, often very coarse, in countries that use them. Electron microscopy (Fig. 4.10) confirms the large particle size and shows that the composition is mixed. Subsequent analysis of the X-ray signal showed that most particles were alumino-silicate in composition. As discussed above, the soiling that is modelled in the current studies relates to small elemental carbon particles, mainly deriving from traffic emissions. The latter type are almost certainly present in the sample but are difficult to separate out from the dominant tyre and road dust particles. Examination of other outliers from Fig. 4.8 revealed different particle types. In Antwerp, for example an unusual iron rich particle (orange coloured in the light microscope) was visible in a number of specimens, while in Krakow a number of industrial particles were seen – almost certainly emanating from the Nova Huta steelworks at the outskirts of the city.

These few illustrations demonstrate that particle assemblages are complex and that dose-response functions need to be reviewed in the light of any situations particular to the site in question. It is certainly prudent to treat the soiling dose-response functions with due caution since, as demonstrated, they refer to a complex mixture of particles that changes from location to location.

4.6 Dose-Response Functions for Soiling

The mathematical theories for the growth of particulate matter on the surface experiencing soiling predict an exponential relationship with the soiling constant k directly proportional to the amount of deposited particulate matter and hence to the atmospheric concentration C, if the resuspension/removal rate Y is negligible. In these circumstances, k is directly proportional to the particulate pollution concentration C.

Substituting into Equation (4.1) yields the generic dose-response function

$$R = R_0 \exp(-\lambda Ct) \tag{4.2}$$

where λ is the Dose-Response constant.

A series of experiments conducted in London, Krakow and Athens under the MULTI-ASSESS project allowed this theoretical relationship to be tested against experimental data. From the resulting fit of experimental data against the physical model, dose-response functions were established.

4.6.1 Loss in Reflectance and Soiling Constant k

Figure 4.11 shows the result of fitting equation (4.1) to data for white painted steel at London Marylebone Road.

The soiling constant for this set of data was $k = 0.000398$/day. The analysis was repeated for the other sites and soiling constants for steel, plastic, membrane and limestone were calculated.

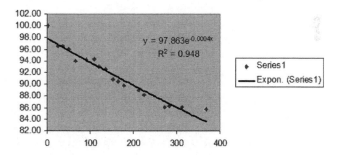

Fig. 4.11 Marylebone Road (London); soiling data fitted to an exponential curve

4.6.2 k vs. PM₁₀ and the Dose-Response Constant λ

The results allow the dependence of the soiling constant k on PM_{10} concentration to be assessed. According to Equation (4.2), the relationship should be linear, if the resuspension/removal rate is negligible. Results are shown in Fig. 4.12.

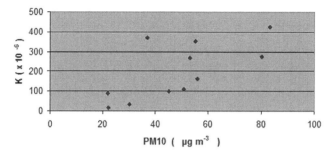

Fig. 4.12 The relationship between soiling constant and PM_{10} for white painted steel

4.7 Dose-Response Functions

Applying a best linear fit to the results shown in Fig. 4.12 has allowed the soiling constant k for white painted steel to be related to the PM_{10} concentration. A similar methodology was applied to the other materials. Substituting these relationships into the basic exponential model, results in dose-response functions as follows (Watt, Jarrett and Hamilton, 2008):

- Painted Steel
  $$\Delta R = Ro\,[1\text{-}\exp(-\,C_{PM10} \times t \times 3.96 \times 10^{-6})]$$
- White Plastic
  $$\Delta R = Ro\,[1\text{-}\exp(-C_{PM10} \times t \times 4.43 \times 10^{-6})]$$
- Polycarbonate Membrane (a convenient indicator material)
  $$\Delta R = Ro\,[1\text{-}\exp(-C_{PM10} \times t \times 3.47 \times 10^{-6})]$$
  where: C_{PM10} is in $\mu g\ m^{-3}$; t is in days
  The results for limestone revealed too much scatter to predict a relationship with any confidence.

4.7.1 The Use of the Dose-Response Functions

The dose-response functions for the soiling of materials have been used for three applications; predictions of soiling trends with time, maps showing spatial variations in soiling and establishing an air quality standard based on soiling.

The variation of soiling with time can be estimated by substituting appropriate values of t into the dose-response functions listed above. For example, for white painted steel, the equation is

$$R = Ro[1 - \exp(-C_{PM10} \times t \times 3.96 \times 10^{-6})]$$

For a typical site with $C_{PM10} = 40\ \mu g\ m^{-3}$, the variation in soiling experienced by white painted steel is shown in Fig. 4.13. There is, of course, a

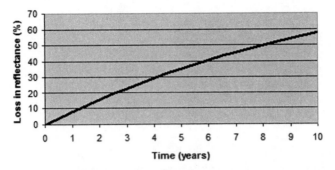

Fig. 4.13 Variation in soiling with time, white painted steel exposed to $C_{PM10} = 40\ \mu g\ m^{-3}$

significant uncertainty in extrapolating beyond the time frame used to develop the model and this uncertainty will only be reduced when a database covering a much longer time frame becomes available, as is the case with the corrosion results reported in Chapter 3.

The Dose-Response function can be combined with a pollution map of a region to produce a map showing the predicted soiling rate of materials in that region. Figure 4.14 shows such a map for white painted steel in London.

These Dose-Response functions can be used to predict the 1-year loss in reflectance as a function of the ambient concentration, C_{PM10}, to which the material is exposed. The result for white painted steel is shown in Fig. 4.15 below:

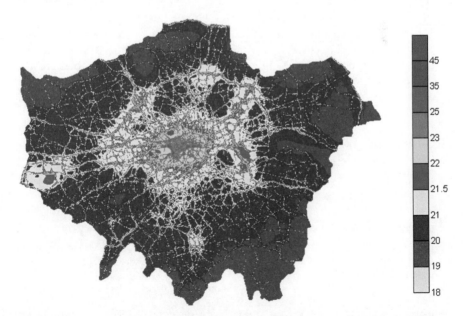

Fig. 4.14 A soiling map for London; white painted steel, % loss in reflectance after 5 years

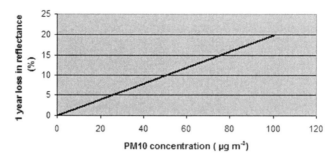

Fig. 4.15 Variation of soiling with PM_{10} concentration (white painted steel)

This information could, in principle, be combined with knowledge of an acceptable or tolerable level of soiling to establish the appropriate air quality standard based on the soiling of materials. The task of identifying an acceptable or tolerable level of soiling relies on detailed knowledge of two factors:

1. The public attitudes to soiling and a perception of what constitutes acceptable degradation have indicated that a 35% loss in reflectance triggers significant adverse public reaction.
2. An assessment of the period of time for which the building can remain without cleaning and an economic evaluation of the options.

These factors could, in principle, be combined to make an estimate of the maximum ambient air quality to which the building can be exposed before the level of soiling becomes unacceptable. This would allow the soiling of buildings to be included as a factor when conducting an environmental assessment for a proposed development, such as the implementation of a major traffic scheme. One study has reported the changing patterns of soiling and microbial growth on building stone in Oxford, England, following the implementation of a major traffic scheme (Thornbush and Viles, 2006). The implications of this approach will be discussed further in Chapter 9.

Acknowledgments The authors of this chapter wish to express their appreciation to all the colleagues in the MULTI-ASSESS and CULT-STRAT consortia for their support, provision of data and stimulating discussions.

References

Aksu R, Horvath H, Kaller W, Lahounik S, Pesava P and Toprak S (1996) Measurements of the deposition velocity of particulate matter to building surfaces in the atmosphere. *J Aerosol Sci.* 27, S675–676

Beloin NJ and Haynie FH (1975) Soiling of building materials. *J Air Pollut Control Ass.* 25, 393–403.

Brimblecombe P and Grossi CM (2005) Aesthetic thresholds and blackening of stone buildings. *Sci Total Env.* 349, 175–198.

Creighton NP et al., (1990) Soiling by atmospheric aerosols in an urban industrial area. *J Air Waste Manag Assoc*. 40, 1285–1289.

Davidson CI, Tang F, Finger S, Etyemezian V and Sherwood SI (2000) Soiling patterns on a tall limestone building: changes over 60 years. *Environ Sci Technol*. 34, 560–565.

Del Monte M, Sabbioni C and Vittori O (1981) Airborne carbon particles and marble. deterioration, *Atmos Env*. 16, 2253–2257.

Grossi CM and Brimblecombe P (2004) Aesthetics of simulated soiling patters on architecture. *Environ Sci Technol*. 30, 3971–3976.

Hamilton RS and Mansfield TA (1992) The soiling of materials in the ambient atmosphere. *Atmos Env. Part A – Gen Topics*. 26, 3291–3296.

Haynie FH (1986) Theoretical model of soiling of surfaces by airborne particles in aerosols: research, risk assessment and control strategies. In *Proceedings of the Second US-Dutch Symposium*, ed. Lee SD, Lewis Publishers, Williamsburgh, VA.

Haynie FH and Lemmons TJ (1990). Particulate matter soiling of exterior paints at a rural site, *Aerosol Sci Technol*. 13, 353–367.

Horvath H (1993) Atmospheric light absorption – a review. *Atmos Env. Part A – Gen Topics* 27, 293–317.

Ionescu A, Lefèvre RA, Chabas A, Lombardo T, Ausset P, Candau Y, Rosseman L. (2006). Soiling modelling based on silica-soda-lime glass exposure at six European sites, *Sci Total Environ*. 369, 246–255.

Lombardo T, Ionescu A, Lefèvre RA, Chabas A, Ausset P and Cachier H., 2005. Soiling of silica-soda-lime float glass in urban environment: measurements and modelling, *Atmos Environ*. 39, 989–997.

Lanting RW (1986). Black smoke and soiling in aerosols: research, risk assessment and control strategies. In *Proceedings of the Second US-Dutch Symposium*, Ed. Lee SD, Lewis Publishers, Williamsburg, VA.

Mansfield TA and Hamilton RS (1989). The soiling of materials: models and measurements in a road tunnel. In *Proceedings of the 8th World Clean Air Congress*. Eds Brasser LJ and Mulder WC. Elsevier Science Publishers B.V., Amsterdam, The Netherlands, 353–357.

MULTI-ASSESS (2007) Model for multi-pollutant impact and assessment of threshold levels for cultural heritage. EU 5FP RTD project contract: EVK4-CT-2001-00044.

Newby PT, Mansfield TA and Hamilton RS (1991) Sources and Economic Implications of Building Soiling in Urban Areas. *Sci Total Env*. 100, 347–366.

Parker A. (1955) The destructive effect of air pollution on materials. National Smoke Abatement Society. London. pp 3–15.

Pesava P, Aksu R, Toprak S, Horvath H and Seidl S (1999) Dry deposition of particles to building surfaces and soiling. *Sci Total Env*. 235, 25–35.

Pio CA et al., (1998) Atmospheric aerosol and soiling of external surfaces in an urban environment. *Atmos Env*. 32, 1979–1989.

Saiz Jimenez C (1993) Deposition of airborne organic pollutants on historic buildings. *Atmos Env*. 27B, 239–251.

Spence JW and Haynie FH (1972) *Paint technology and air pollution: a survey and economic assessment* AP-103 U.S. Environmental Protection Agency, Office of Air Pollution Research, Triangle Park, NC, USA.

Swedish Corrosion Institute (KIMAB) (2006) http://www.corr-institute.se/MULTI-ASSESS.

Thornbush M and Viles H (2006) Changing patterns of soiling and microbial growth on building stone in Oxford, England after implementation of a major traffic scheme *Sci Total Env*. 367, 203–211.

Toprak S, Aksu R, Pesava P and Horvath H (1997) The soiling of materials under simulated atmospheric conditions in a wind tunnel. *J Aerosol Sci*. 28, S585-586.

Watt J. (1998) Automated Characterisation of Individual Carbonaceous Fly Ash Particles by
      Computer Controlled Scanning Electron Microscopy -Analytical Methods and Critical
      Review of Alternative Techniques. *Water AirSoil Pollut.* 106, 309–327.
Watt J, Jarrett D and Hamilton R (2008) Dose-Response functions for the soiling of materials
      due to air pollution exposure. *Sci Total Env.* 400, 415–424.

Sources of Additional Information

Hinds WC (1999) Aerosol Technology: Properties, behaviour and measurements of airborne
      particles, 2nd edition. Wiley ISBN 978-0-471-19410-1.
Verhoef LGW (1988) *Soiling and cleaning of building façades.* Taylor & Francis. ISBN
      0412306700.
Watt J and Hamilton RS (2003) Soiling of buildings by particulate matter. In *The Effects of
      Air 'Pollution on the Built Environment.* Ed Brimblecombe P. Imperial College Press.
      London.

Chapter 5
Some Aspects of Biological Weathering and Air Pollution

Wolfgang Krumbein and Anna Gorbushina

5.1 Overview

The influences of air pollution and deterioration by microorganisms on build-ings, monuments, and sculptures made of stone gained considerable attention in the last Century. Both damage functions were usually discussed separately. In this contribution we deal with the mutual interactions of both factors on stone monuments, which is an emerging area of research.

Practically no material exposed to surface conditions is capable of perma-nently resisting biological settlement, growth and associated transformation. Materials submersed in water, exposed to outdoor or indoor atmospheres and even those conditioned in protective envelopes or gases invariably – with time passing – show traces of the attachment, settlement, invasion and transformation by organisms and their metabolic products. The interactions are multiple and complex, may shorten (or even prolong) the existence of the surface in question, and may change all the physical and chemical aspects of the original material.

An everyday example that may help some of the effects to be envisaged is by thinking of looking at the glass window of an aquarium and seeing the growth of algal or bacterial biofilms at the side exposed to water. Usually near the water level this biofilm is markedly thicker, and it may also grow on parts exposed to the inner atmosphere of the aquarium, where there is humidity creeping up or water splashed up by an aeration unit. Few aquarium owners, however, would think that an invisible biofilm composed of different, albeit invisible, micro-organisms is also growing at the outer glass surface exposed to the visitor side of the aquarium. Fat, dust, humidity and volatile components enable attachment and growth of microbes, which eventually might etch the glass surface to make it as clouded as an old stable window. The aquarium owner thus will (1) add some substances to the water, which will kill or prevent from attachment some of the algal or bacterial components without causing harm to the beautiful fish,

W. Krumbein (✉)
University of Oldenburg, Ammerländer Heerstraße 114-118, D-26129 Oldenburg, Germany
e-mail: wek@uni-oldenburg.de

and (2) from time to time clean carefully both sides of the aquarium window. Of course there is a great deal of variability – laboratory atmospheres differ from outdoor atmospheres and the atmosphere of a living room from that in a public show aquarium with many visitors and often very high room humidity but this simple example suggests a lot for biodeterioration of materials and its prevention.

This chapter will examine some of the questions posed. It will deal to some depth with the individual organisms and their activities.

In general terms, biodeterioration has to do with general phenomena and processes related to ageing.

5.2 Introduction

Krumbein and Dyer (1985) suggested separating the process of material destruction into physical and chemical transfer reactions. Physical transfer includes biogenic and abiogenic processes through which particles are mechanically removed from any material and transferred into the hydrosphere, atmosphere or pedosphere as particulates, colloids and aerosols. Chemical transfer embraces all chemical reactions -biotic and abiotic- through which materials are removed and transferred into the form of gases, solutes, colloids and particulates (chemical reactions can yield solid particles after a chemical transfer and chemical reaction ultimately yielding precipitates after an intermediate gaseous or liquid state).

Krumbein (1988) described the physical and chemical biotransfer actions in an embracing definition as:

Biocorrosion, biodeterioration, bioerosion or bioweathering is an exchange or biotransfer of material and energy between two heterogeneous open systems: a usually solid substrate (rock, concrete, mural painting, glass, wood, leather, parchment paper etc.) and its environment (mainly atmosphere and hydrosphere). Both systems are defined by their physical properties (mass, volume, humidity, pressure, porosity etc.), chemical composition, biological components (physiological, biochemical, biophysical) and through their innate and permanently added or subtracted energy. The mutual interaction of all components and processes leads to a more or less complete turnover of the initial materials at the border between the two systems. These activities are limited by the penetration depth of physical gradients, gases, solutions, and organisms into the material. This (bio-) transfer process may come to a standstill for certain periods of time when the conditions approach equilibrium (e.g. through patina, crusts etc.). It may, however, be revived only if one of the components or processes involved changes or is submitted to changes.

Krumbein and Dyer (1985) and Krumbein (1988) have tried to define and explain the terms frequently in use in this field. Weathering with its meteorological connotation is a somewhat awkward term not directly implying the important biological interactions in the process of rock and material decay. It should therefore be avoided in material sciences when biological processes are involved. Erosion and abrasion involve the physical attack of solids, liquids or gases on solids with the effect of detaching and transporting particles. Corrosion implies a chemical transfer with consequential surface changes. It is used in context with minerals and especially with metals and alloys and to a lesser extent with stone or other materials. Degradation is a term for the combination of weathering, surface destruction and erosion; it is also frequently used for the biological breakdown of organic or inorganic compounds which serve as energy, electron or nutrient source for the organism. Also the formation of new mineral lattices and chemical compounds (rust) may be involved. Deterioration and biodeterioration are terms widely used in material sciences.

5.3 Previous Work and Questions Posed

There is a growing concern that ecosystems (including big cities and their cultural heritage) respond to anthropogenic changes of environment not only by gradual changes, but also by abrupt stage shifts. The new, sometimes undesirable alteration may be difficult to reverse, even if environmental pressure is relaxed. Such persistent changes in biological communities or ecosystems are frequently used for bioindication purposes in assessment of anthropogenic impact. In urbanised areas biological ecosystems are drastically reduced in their natural scope and diversity, and for bioindication purposes ubiquitously present biological communities are selected. Microscopic growth accumulations dominated by micro-organisms (bacteria, fungi, algae and lichens) are universally present on all solid/air interfaces and are referred to as "subaerial biofilms". The subaerial biofilm microscopic diversity represents an equilibrated multi-component open ecosystem sensitively reacting to all environmental factors (Gorbushina, 2007).

All parts of these diverse communities react very sensitively to such external changes as (i) microclimatic conditions (including mean temperature rise as a consequence of global climate change) and (ii) organic deposition from the atmosphere. These changes can be sensitively monitored and analysed using up-to-date microbiological, molecular biological, biotechnological, and biophysical methods. The bioindicative potential of these unique microbial communities until now was grossly underestimated, as only lichenoindication was previously employed (Owczarek et al., 1999). Special concern has to be given to the changes as they occur in the atmosphere of big cities, where on the one hand, an increased level of pollution is present and on the other hand, a considerable part of our built cultural heritage is concentrated. High levels of organic pollution in big European cities make classical lichenoindication impossible, because lichens are frequently not able to cope with such environmental impacts. Data exist, however, that suggest growth of subaerial biofilms can even be stimulated by particulate and volatile organic input from the atmosphere. As shown by Krumbein (1966) particulate and volatile organics, adsorbed to rock surfaces may drastically increase the growth of subaerial biofilms on monument surfaces. Such changes have been analysed in detail by Machill et al. (1997), and Saiz-Jimenez (1995).

Evidence exists that biofilm or biopatina growth could be used as an efficient and powerful indicator of climate change (Gorbushina, 2007). For example, drastic changes in biological diversity observed on rock surfaces in urban and rural environments (Krumbein, 1966; Sterflinger and Prillinger, 2001) are directly connected to differences in air quality, especially regarding anthropogenic input of organic substances (Krumbein and Gorbushina, 1996). These changes in biological diversity, in their turn, have a direct impact on weathering processes as they take place on building and monument surfaces. It is firmly established that microbial biofilms on stone surfaces lead to a change in the colour of the rock surface and more severe erosion phenomena as was convincingly shown for the Acropolis monuments in Athens (Urzì et al., 1992; Viles and Gorbushina, 2003). All these biogenic phenomena at the rock/atmosphere interface are under the direct influence of the atmospheric input. Rock biofilms are ecologically sensitive systems (Gorbushina, 2007). In the present situation of climate change and increased nitrogen and organic pollution mainly as a result of traffic emissions these indicator communities are exhibiting increased levels of undesirable alterations. Since the latter almost always are expressed by spectral alterations of the surfaces colonized by biofilms it is important to relate such colour shifts to the established biofilm communities and their changes with time and especially increased levels of physical and chemical pollution. Modern physical equipment allows to measure or define (1) the physical or climate environment and pollution and (2) the response of subaerial biofilms to the impact of nitrogen and especially organic pollution by estimating the shift in autotroph (algal and cyanobacterial) and heterotroph (bacteria, actinobacteria, fungi) biofilms of indoor and outdoor movable and immovable cultural heritage objects.

5.4 Materials and Environment – Air Pollution and Microbes

Practically all materials are vulnerable to biological attachment and growth, and subsequent material destruction or biodeterioration. The decision to treat this biodeterioration is largely dependent on what the material is used for and the significance attached to this. For example, materials used frequently in buildings and construction, transport vehicles and objects of art are seen as more valuable and are therefore subject to more intensive levels of research and restoration.

Which material will endure best under which atmospheric conditions is often uncertain and needs careful consideration and examination of previous exposures. Often very astonishing and unexpected observations need to be

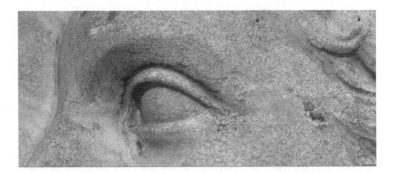

Fig. 5.1 John Keats Epitaph near the Cestius pyramid in Rome shows biological pitting destruction just under the eyelid where usually tears drop down. In this case humidity of rain collecting below the eye initiated biological growth and destruction (see Fig. 5.2)

explained. Marble of high sculpture quality used for statues in the park of Sanssouci (Potsdam) suddenly crumbled into sand, short after transfer into "protective" museum atmosphere, while some of the finest sculptural elements of the marble freeze of the Parthenon (Athens) lasted for almost 3,000 years under outdoor atmospheric conditions until they were recently transferred into

Fig. 5.2 Microcolony of black rock eating yeast like fungus etching its way into Pentelic marble. Roundish cells form a protected cluster and etch their way into the rock creating biopits as seen in Fig. 5.1 (scale bar 5 micrometre)

the restorers studio for cleaning and transfer to the Acropolis Museum. Future generations will have to see the outcome of this move aimed at protection. Another example is a small Romanesque church in Austria, which contains a sandstone-stone Altar table on which a lime-wood crucifix has been positioned for several hundreds years. The wooden crucifix is as lovely and perfect as when it left the artist's studio, while the stone supporting it has been replaced several times already. In other cases fire, insects and fungi have caused loss of an oak-wood roof construction above a castle made of granite. The roof had to be replaced several times, while the stone construction lasted on.

The creation of a sequence of materials concerning increasing resistance to biological growth and degradation is difficult to judge without considering environmental influence, type of organism and surface treatment of the material. A scale of resistance (like the Mohs scale of hardness of minerals) is practically impossible to give. Long experience in microbial deterioration of materials indicates, however, that organic substances may last as long as inorganic ones. Some rocks are very resistant (e.g. porphyry, special basalts and granites as well as very dense limestones and marbles), while others tend to be extremely vulnerable (e.g. most sandstones if not fully quartzitic, many marbles, some granites). Some kinds of wood are extremely resistant (lime-wood, apple, mostly the so-called hardwoods), others are rapidly decayed by fungal rot (softwood like pine, fir, poplar). The example of the Dead Sea Scrolls shows the resistance of papyrus being high in comparison to e.g. copper scrolls. Pottery, Adobe and other artificial building materials are more resistant than many natural rocks, while many types of concrete are very vulnerable to acid producing bacteria. Attempts of a biodeterioration scale were published but without really quantitative field or laboratory evidence (Griffin et al., 1991; Gorbushina and Krumbein, 2006; Kondratyeva et al., 2006).

5.5 Types of Organisms and Their Deteriorative Potential

Organisms, to exist and to metabolise, need water. Water supply even in minimal amounts is therefore essential. Survival units, however, may last thousands of years before germinating and metabolising again. All types of organisms have been reported to cause damage to construction materials. Cemeteries, old castles and fortifications such as the Great Chinese Wall, the monuments of Angkor Vat, Borobodur, Middle America, the Roman Aqueducts, and the Coliseum in Rome are good examples for the influences of higher plant vegetation impact (Figs. 5.3 and 5.4). Interactions with animals are less evident. However many types of insects are active in material decay (termites, mites, singular wasps and bees, even spiders or birds may act on rock and other building materials (Gorbushina and Petersen, 2000).

Fig. 5.3 Large tree roots deforming and crushing parts of one of the Temples in the Sanctuary of Angkor Vat (Cambodia). Mosses and smaller plants penetrate into crevices and cracks enlarging them and pushing the stones apart

Fig. 5.4 Tree growing over and into tombstones on the old Jewish cemetery of Prague (Czech republic)

Today, however, a general opinion has emerged, that several groups of micro-organisms are more significant than others in biodeteriorative processes of the built environment.

5.5.1 Macroscopically Visible Organisms

Besides the generally limited effects of trees and roots on buildings or tomb-stones e.g. at Angkor Wat and on the Prague Jewish cemetery (Figs. 5.3, 5.4) mosses, Lichen and Algae are macroscopically evident because they cover the material surfaces with visible films of growth. Consequently an aesthetic altera-tion is first noticed. In addition they have a corrosive effect on the substrate by release of acid metabolites, some of which are chelating agents that determine a solubilising and disintegrative action on the constituents (Chen et al., 2000). Surficial microbial mats, films and lichen colonies by their exudates (mainly sugars) act as traps of dust and particles, which in turn supply aggressive compounds and nutrients for all kinds of organisms. Lichens provoke physical as well as chemical damage through the formation of pits or crater shaped holes, and surface disintegration and solubilisation by acid metabolites (Figs. 5.1, 5.2). Pits and craters are produced in some cases (epilithic lichen) by the algal symbiont, in other cases (mainly endolithic lichen) by the fruiting bodies of the fungal symbiont (Gehrmann and Krumbein, 1994). It is of utmost impor-tance to note that some lichen can act protectively, while others on the same rock act destructively. Therefore cleaning action has to be considered carefully. Many algae form unaesthetic growth films, slime, which upon their degradation yield biocorrosive actions and can actively perforate the rock.

5.5.2 Microorganisms in General

It is well known that microorganisms are involved in rock and mineral decay. A correlation between specific stone decay patterns and the presence of specific microorganisms presents some difficulties because a great number and variety of species of microorganisms are involved. In fact, many autotrophic and heterotrophic bacteria, algae and fungi are found on material surfaces but a biodeteriorating role has been demonstrated only for some of them. In many cases microorganisms are directly associated to a deteriorating activity on materials; in others products excreted from the cells under stress or upon lysis serve as nutrients for other heterotrophic decaying microorganisms or act directly on the material without a direct correlation to the organisms. Micro-organisms are defined as organisms barely or not visible with the naked eye. Microorganisms can belong to the animals, the plants, the fungi, the algae, the protozoans (protoctists) and to the prokaryotes (bacteria sensu latu).

5.5.3 Phototrophic Microorganisms

Lichen and Cyanobacteria, as predominant rock dwellers are usually present in association with green and red algae and diatoms. The dominance of one or other of these groups varies both locally and regionally. Cyanobacteria (blue-green algae) and green or red algae growing on surfaces within such buildings as churches are well adapted to survive at very low light levels. Community colour varies with the colour and growth form of the dominant forms (John, 1988). Cyanobacteria are frequent in the tropical and arid areas where they thrive

Fig. 5.5 Black biofilm on specific places in the restored backside of the Erechtheion (Akropolis, Athens, Greece). Ancient and replaced marble blocks are recognised by different biopatina. Black biofilm spreads in places where – behind the wall – water can pass

Fig. 5.6 Wall of the "Neue Pinakothek" München (Germany). Water runs down the sandstone plates and causes black biofilms also called "Tintenstrich-Flora" (ink-streak flora). Black soot or any air pollution particle can be excluded as the source of damage. The "Neue Pinakothek" was cleaned over a period of several years and a roof protection protruding only 3 cm horizontally from the rooftop was added contrary to the initial design of the architect according to whom the stones should directly meet the sky above

(a)   (b)

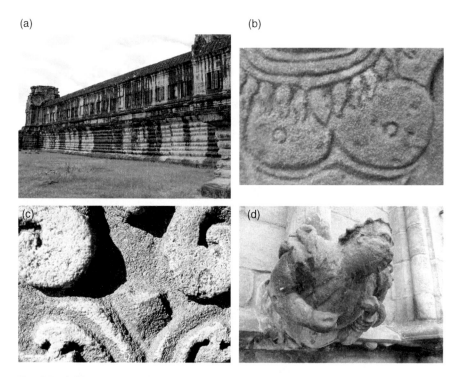

(c)   (d)

Fig. 5.7 Biofilm cover at different sites. *Upper left*: Black biofilms composed of algae, cyanobacteria, and fungi at Angkor Wat; *upper right*: Angkor Wat Temple freeze with *red* pigment of lichen growth and biopitting fungi; *lower left*: *blue* biofilm pigment from lichen growth in Thai temple, Southern Thailand; *lower right*: *black* and *green* biofilm on lion heads at Sterling Castle, Scotland. In all cases neither pollution nor artificial paint traces exist, it is pure biological growth

either at the surface or up to several millimeters deep. This is due in part to high temperatures and/or extremely high or low humidity as well as high or low irradiation regulating the points of occurrence and attack.

Cyanobacteria and algae, like fungi, can form biofilms and crusts on rock surfaces that are deep or bright green in humid conditions and deep black when dry – called ink-streack flora (Figs. 5.5, 5.6, 5.7, 5.8). The black coatings of many rock surfaces can be partially explained this way. Apart from the evident aesthetic damage on stone monuments there is evidence of significant physical and chemical deterioration of the surface by excretion of chelating organic acids and sugar derived carbonic acids, which initiate the perforating activities of cyanobacterial, activity that essentially breaks down the stone and opens up fissures and crevices. Some of them are called endolithic, when they exhibit strong material perforating activity. Algae are eukaryotic phototrophs and can also act in this way; especially in connection with fungi or as symbiont of lichens.

Fig. 5.8 Actively growing lichen on a Jewish tombstone (Southern Germany). The lichen may leave behind biofilm colours (organic pigments) in *yellow, red, blue, grey and green* spectral appearance

5.5.4 Chemolithotrophic Microorganisms

Sulphur compound oxidizers. High numbers of strictly autotrophic (i.e. capable of synthesizing its own food from simple organic substances) *Thiobacillus* sp. have been found not only under the surface of highly deteriorated stones, which presented a pulverizing aspect, but also on deeper layers (10 cm) where there was no stone decay yet visible. Some anaerobic species, like *Desulfovibrio desulfuricans*, are not strictly autotrophic and can sometimes utilise organic compounds as electron donors. They can find sulphate from air pollutants or from soil and produce hydrogen sulphide as follows:

$$\text{Organic acid} + SO_4 \rightarrow H_2S + CO_2$$

This product is highly corrosive acid and its salts provoke on open air stone surface the formation of black and/or grey films and crusts often described as patina. Sulphate-reducing bacteria increase the decay activity of sulphur and sulphide or thiosulphate—oxidizing bacteria causing a deeper biodeteriorating action. In fact, on soil *Desulfovibrio desulfuricans* reduces sulphates to sulphites, thiosulphates and sulphur. By capillarity these compounds can reach the superficial layers of stones where sulphur-oxidizing bacteria will oxidize them to sulphuric acid. This strong acid reacts with calcium carbonate to form calcium sulphate (gypsum), which is more soluble in water than calcium carbonate (calcite, aragonite and dolomite).

Nitrifying bacteria. The nitrifying bacteria are commonly found on deteriorated surfaces of rock materials. Dissolved and particulate ammonia is deposited on rock surfaces by rain and wind from various sources, among which traffic and agricultural sources (fertilizer, manure) are the most influential. Bird excrements and other ammonia and nitrite sources are also oxidized microbially by chemolithotrophic and in part, also by heterotrophic ammonia oxidizers and nitrite oxidizers, to nitrous and nitric acid.

This transformation is divided into two steps:

1) OXIDATION OF AMMONIUM by Nitrosomonas, Nitrosococcus, Nitrosovibrio, Nitrosospira:

$$NH_4^+ + \frac{3}{2}O_2 \rightarrow 2H^+ + NO_2^- + H_2O$$

2) OXIDATION OF NITRITE by Nitrobacter, Nitrococcus, Nitrospira

$$NO_2^- + \frac{1}{2}O_2 \rightarrow NO_3$$

Both of the resulting acids attack calcium carbonate and other minerals. The CO_2 produced can be utilised by them to form organic compounds while calcium cations form nitrates and nitrites, again more soluble than the original mineral phases. Capillary zones can accumulate these products and hydrated and non-hydrated forms of such salts can create considerable damage by volume changes of the mineral phases. The characteristic symptom of the activity of nitrifying bacteria is a change of stone properties. The rock becomes porous, exfoliation occurs and fine powder may fall off, which sometimes is yellow from freshly formed iron oxides. Some evidence of organotrophic bacteria causing nitrification has also been collected (Bock pers. comm.).

5.5.5 Iron and Manganese Oxidizing Microorganisms

The most common iron oxidizing microorganisms which are not living exclusively in lakes and flowing water like Gallionella or Siderococci belong to the groups of fungi, chemoorganotrophic bacteria (Arthrobacter) or to the autotrophic group like *Thiobacillus ferrooxidans*, *Ferrobacillus ferrooxidans*, *Ferrobacillus sulfoxidans*. The type *Metallogenium symbioticum* has often been reported also in connection to weathering. Krumbein and Jens (1981) have isolated numerous Fe- and Mn-precipitating microorganisms from rock varnishes and some of the isolates closely resembled to *Metallogenium symbioticum*. Iron-oxidizing bacteria directly attack iron-bearing rocks as well as any structure made of iron associated with stone monuments.

5.5.6 Chemoorganothrophic Microorganisms

In recent years a constantly growing number of contributions have been made about the impact of chemoorganotrophic microbiota on the deterioration and biotransfer of inorganic materials with no direct source of organic substrate. Krumbein (1966) showed convincingly that the organic pollution within large cities may increase the abundance of chemoorganotrophic bacteria in rocks transferred from rural environments by a factor of 104 within one year. This group of microorganisms needs organic compounds as energy and carbon sources. Stone (as an inorganic material) can support chemoorganotrophic processes for several reasons.

Four different types of sources of organic materials are usually present in variable concentrations on rock and mineral surfaces (mural paintings especially). These are:

1. Consolidating and improving products applied to the surface such as wax, casein, natural resins, albumin, consolidants, hydrophobic agents.
2. Dust and other atmospheric particles (aerosols) and organic vapours from anthropogenic sources. The latter consist mainly of hydrocarbons from aircraft and car fuels or power plants but also of agricultural applications such as manure and harvesting dusts or just pollen and other compounds excreted into the air by plants and animals. In industrial areas and cities, organic sources stem from industry and crafts such as food factories, bakeries etc. Recently it has been calculated that the total amount of fossil fuels in terms of organic carbon could derive from the annual production of pollen and etheric oils and volatile etheric substances (smell of flowers) that are carried into the sea.
3. No less important is the coexistence of photo- and chemolitothrophic microorganisms with chemoorganothrophic ones in biofilms, biological patina and microbial mats encrusting the upper parts of rocks or the whole paint and mortar layer of frescoes. In this context the chemoorganotrophic bacteria and fungi can survive and reproduce easily because they have all nutrients from autothrophic organisms.
4. Sedimentary rocks usually contain between 0.2 and up to 2% organic matter retained in the rock. These compounds serve as source of energy and carbon for many microorganisms.

Thus practically all groups of chemoorganotrophic bacteria may live on and in rock materials and will invariably increase in numbers and activities with increasing levels of pollution.

5.5.6.1 Fungi & Actinomycetes

Fungi are very commonly found on stone surfaces. Their deteriorating effect is due to mechanical and chemical action (Jones and Wilson, 1985; Jongmans et al., 1997; Dornieden et al., 2000). The first one is related to hyphae penetration of materials that has deep reaching deteriorating effects such as swelling and deflation as physical effects, channelling water into and keeping it in considerable depths and constant microvibrations through micromotility; the second one is due to production of acid metabolites (oxalic, acetic, citric and other carbonic acids). The latter have a chelating and solubilising effect on many minerals. New findings and reports hint to the fact, that similar to the activity in mineralization of organic matter in soil, fungi may also be the most important rock and material dwellers and transformers. This role they may have taken already in the late Precambrian with deep impact on all element cycles in nature (Gorbushina, 2007).

Actinomycetes are so called because of their similarity to fungi. Among these microorganisms, the genus *Streptomyces* is most frequently occurring in biodeteriorated stone. Penetration of the substrate by their hyphae is increased by their ability to excrete a wide range of enzymes. They can form a whitish veil or a granulose patina on mural paintings. They also produce several water soluble dark pigments. Recently it was documented, however, that they rarely if ever produce noteworthy amounts of organic acids and chelates in a rock decay environment. They may have less detrimental activities. Their presence in high numbers, however, indicates a very intense population of fungi and other microorganisms. They can perfectly well resist to dryness and therefore can be regarded as excellent indicators of a progressed infection of rocks and mural paintings by other microorganisms. Early stages of rock infections usually do not exhibit high numbers of this group of actinomycetes. The coryneform and nocardioform bacteria, however, have been identified as frequently occurring on rock materials and producing acids from organic pollutants on building stones thus contributing considerably to biocorrosion and biodeterioration, and these groups have recently been placed into the actinomycetes group (Goodfellow et al., 2009).

5.6 Interactions of Air Pollution and Biological Weathering

Many components are released to the atmosphere in large amounts by natural and anthropogenic means. They are often summarized as air pollutants, which is not always true. These are (without implying any ranking)

- Sulphur compounds (all species but mainly SO_2)
- Nitrous compounds (NO_x)
- Methane
- Carbon dioxide
- Hydrocarbons
- Carbohydrates
- Phenolic compounds
- Derivatives of natural and industrial polymeric materials
- Particles of different size classes
- Organisms and parts of organisms (e.g. bacteria, fungi, spores, pollen)

All of these compounds and particles may interact with material surfaces directly and cause different kinds of damage (Kucera, 2003). In this chapter, however, we only treat the indirect influence of these compounds and particles on microbial growth on materials exposed to the atmosphere. If we correctly regard the microbial communities on buildings and natural rock surfaces as successions of growth of several different and interacting microbial groups and species and summarise them as subaerial biofilms (Gorbushina, 2007) as contrasted to sub aquatic ones (Costerton et al., 1995) it has to be considered, how such communities or biofilm systems interact with input from the atmosphere. Subaerial biofilms usually retain water and humidity originating from rain, fog or runoff from roofs and walls for long periods of time. The microbes within a biofilm community also are often embedded in large amounts of extra cellular polymeric substances of a slimy and sticky nature (EPS). The stickiness and glue-like behaviour of such biofilms is most drastically expressed in horizontally oriented sandstone plates of terraces or sidewalks which can get extremely slippery. These biofilms may prevent direct diffusion of potentially aggressive acid anhydrides (SO_2, NOx etc.) into the materials. The biofilm community may use these compounds as electron acceptors or donors and even as energy source (e.g. $S_2O_3^{2-}$). Hereby acid rain can be neutralized.

Methane, carbon dioxide, organic compounds and particles of all classes may in turn be trapped by the biofilm surfaces and used as sources of energy, nutrients and electron donors or acceptors for metabolic pathways. This enables the microbial biofilm, which cannot find much food or nutrients within the material to which it adheres, to thrive better and thus penetrate into the material. The process of physical and chemical penetration into rock fissures, cracks and crevices often along grain boundaries as well as various metabolic products of acidic or highly alkaline character will be enforced by this supply from the atmospheric load. In this way, biodeterioration can reach higher intensities.

Krumbein (1966), and Gorbushina and Krumbein (2004) have elaborated on these interactive forces between atmospheric organic pollutants and heterotrophic growth of bacteria and fungi, which is often logarithmically higher in city centres than in a more rural environment. What can be stated clearly is that most discolorations on rock surfaces are caused at least in part by microbial biofilms. Only in very rare cases are black particles from fly ashes or soot the sole cause (Figs. 5.5, 5.6, 5.7, 5.8).

On the other hand, unlike bacteria and fungi, lichen growth will invariably be suppressed in cities with highly polluted atmospheres (Gehrmann and Krumbein, 1994). A special case was studied between 1985 and 2000 at the Jewish cemetery of Hanover (Germany). This cemetery was situated directly adjacent to a metal processing factory. Only one single species of lichen (Lecanora conizoides) occurred on the tombstones, while in 20 other cemeteries all over Germany up to 25 species were recorded and mapped (Gehrmann, pers. comm., M. Sc. thesis). The factory was closed in 1987. Already in 1997 more then 15 lichen species had returned to the stone surfaces. Lichen may be highly rock deteriorating – especially in the case of Mediterranean marble monuments. Caneva et al., (1995) report biopitting of the Colonna Traiana in Rome and other places while Diakumaku et al., (1995) have pointed out the same phenomenon for the Athens monuments. The biopitting progress at the Acropolis and Agora of Athens is much less intensive and progressive than the natural outcrops of marble on the Philopappos hill nearby, which is submitted to fewer direct traffic derived pollutants or the completely unpolluted, but highly biopitted, Delos sanctuary. Another example of air pollution derived improvement of the biodeteriorative impact is the decrease in damage to mediaeval church windows in some French cities observed by conservators. In the late 19th and early 20th Century lichen growth on stained-glass windows was of considerable importance, whereas bio-pitting and etching has become much less accentuated and problematic over the last 75 years. Indicators for such changes in the interaction between air pollutants and biofilms are the glass sensors developed by the Fraunhofer Institute in Würzburg and tested in several EU projects.

The interaction between lichen and the substrate itself therefore seems much less important than that with the atmospheric load. The only well-established exception from this general observation is the case of several types of crustose lichen (evidently divided into calcicole and silicole). However, this distinction was derived under relatively unpolluted conditions.

5.7 Cost and Benefit of Biological Growth

An especially ambitious task is the economic evaluation of the role of separate (biological, chemical, physical) components in the overall phenomenology of stone weathering/breakdown. Influences of reactions to increased acidity of the atmosphere, to biological impact, and to a global warming of climate may have caused an increase in biological growth on rock surfaces in the past 150 years with a tendency of shifting importance from decreasing air pollution to increasing effects of global climate change (see BIODAM Project, Krumbein et al., 2006).

5.8 Gaps of Knowledge and Future Work

The major gaps in knowledge of the interactive play between mineral and building materials, air pollution and microorganisms have already been mentioned in previous chapters. There is a general lack of quantitative field and/or laboratory data on the acceleration of natural decay and decomposition of rock and other mineral building materials caused by polluting components in the atmosphere and consequently on microbes and their mutual interactions. Several methods of study were published (Hirsch et al., 1995) and some studies tried quantification (Hoffland et al., 2004). Attempts have been made to exclude acid rain and to test aggressive solutions in impregnation and washing cycles in climate chambers. These data, however, can not be transferred to natural environmental conditions. Laboratory experiments with aggressive microorganisms have also been undertaken and showed considerably accelerated rates of decay as compared to sterile blank samples. Field observations followed by laboratory experiments undertaken on marble (Dornieden et al., 2000) have shown that microbial activity may yield irreversible grain expansion and subsequent erosion rates of up to 300 micrometer per annum under laboratory conditions or 2 mm in 100 years in the case of field observations.

Altogether the analysis of the interplay of air pollution with biofilm growth on and in architectural rock surfaces is only at a beginning. Many factors are ill-described and combine with a general lack of quantitative data. Although the cost of eliminating biofilms on buildings is extremely high, research and development of new analytical and protective techniques is ill-defined and lacks funding from national and European sources as well as from the building industry. Future work must concentrate (1) on the understanding of biofilm interaction with pollutants (2) the combined damage functions of pollution and biofilms, (3) on the quantification of the aggressivity of biofilm communities, and (4) on methods and techniques to avoid or eliminate microbial growth at an early stage.

Acknowledgments The authors wish to express their appreciation to colleagues for their support, provision of information and stimulating discussions. We are particularly grateful to Herbert Liedel and Helmut Dollhopf for their permission to use their figure in this chapter.

References

Caneva, G., E. Gori and T. Montefinale (1995). Biodeterioration of Monuments in Relation to Climatic Changes in Rome between 19–20th Centuries. Science of the Total Environment 167: 205–214.

Chen, J., H. P. Blume and L. Beyer (2000). Weathering of rocks induced by lichen colonization – a review. Catena 39(2): 121–146.

Costerton, J. W., Z. Lewandowski, D. E. Caldwell, D. R. Korber and H. M. Lappinscott (1995). Microbial Biofilms. Annual Review of Microbiology 49: 711–745.

Diakumaku, E., A. A. Gorbushina, W. E. Krumbein, L. Panina and S. Soukharjevski (1995). Black fungi in marble and limestones – an aesthetical, chemical and physical problem for the conservation of monuments. Science of The Total Environment 167(1–3): 295–304.

Dornieden, Th., A. A. Gorbushina and W. E. Krumbein (2000) Biodecay of mural paintings as a space/time related ecological situation. Internat. Biodeterioration and Biodegradation 46: 261–270.

Gehrmann, C. K. and W. E. Krumbein (1994). Interaction between epilithic and endolithic lichens and carbonate rocks. III International Symposium on the Conservation of Monuments in the Mediterranean Basin. Venice: 311–316.

Goodfellow, M., P. Kämpfer, H-J. Busse, M. Trujillo, K-I. Suzuki, W. Ludwig and W. B. Whitman (2009). Bergey's manual of systematic bacteriology 2nd edition volume 5. The Actinobacteria, Springer, New York. ISBN 0-387-95042-7

Gorbushina, A. A. (2007) Life on the Rocks. Environmental Microbiology 9: 1613–1631.

Gorbushina A. A. and W. E. Krumbein (2004) Role of organisms in wear down of rocks and minerals. pp. 59–84 In: F. Buscot, Varma, (eds.), Microorganisms in Soils: Roles in Genetics and Functions. Springer, Berlin.

Gorbushina A. A. and W. E. Krumbein (2006) Biological testing of inorganic materials. In: Czichos, C., Smith, H., Saio, L., (eds.) Springer-Handbook of Materials Measurement Methods. Springer, Berlin, pp. 753–768.

Gorbushina A. A. and K. Petersen (2000) Distribution of micro-organisms on ancient wall paintings as related to associated faunal elements. Internat. Biodeterioration and Biodegradation 46: 277–284.

Griffin P. S., N. Indictor and R. J. Koestler (1991). The Biodeterioration of Stone – a Review of Deterioration Mechanisms, Conservation Case-Histories, and Treatment. International Biodeterioration 28(1–4): 187–207.

Hirsch P., F. E. W. Eckhardt and R. J. Palmer (1995). Methods for the Study of Rock-Inhabiting Microorganisms – a Mini Review. Journal of Microbiological Methods 23(2): 143–167.

Hoffland E., T. W. Kuyper, H. Wallander, C. Plassard, A. A. Gorbushina, K. Haselwandter, S. Holmstrom, R. Landeweert, U. S. Lundstrom, A. Rosling, R. Sen, M. M. Smits, P. A. van Hees and N. van Breemen (2004). The role of fungi in weathering. Frontiers in Ecology and the Environment 2(5): 258–264

John D. M. (1988). Algal growths on buildings: a general review and methods of treatment. Biodeterioration Abstracts 83: 81–102.

Jones D. and M. J. Wilson (1985). Chemical Activity of Lichens on Mineral Surfaces – a Review. International Biodeterioration 21(2): 99–104.

Jongmans A. G., N. van Breemen, U. S. Lundstrom, P. A. W. van Hees, R. D. Finlay, M. Srinivasan, T. Unestam, R. Giesler, P. A. Melkerud and M. Olsson (1997). Rock-eating fungi. Nature 389(6652): 682–683.

Kondratyeva I. A., A. A. Gorbushina and A. I. Boikova (2006). Biodeterioration of construction materials. Glass Physics and Chemistry 32(2): 254–256.

Krumbein W. E. 1966. Zur Frage der Gesteinsverwitterung – Über geochemische und mikrobiologische Bereiche der exogenen Dynamik. Ph. D. thesis, Würzburg. 106 p.

Krumbein W. E. (1988). Microbial interactions with mineral materials. In Biodeterioration 7, p. 78–100, edited by D. R. Houghton, R. N. Smith and H. O. W. Eggins, Elsevier Applied Science, London and New York, 843 p.

Krumbein W. E. and Dyer, B. D (1985). This planet is alive - Weathering and Biology, a Multi-faceted Problem. In The Chemistry of Weathering 149, p. 143–160, edited by J. I. Drever, D. Reidel Publishing Company, Dordrecht, 324 p.

Krumbein W. E. and A. A. Gorbushina (1996). Organic pollution and rock decay. p. 277–284 In: Pancella, R. (ed.), Preservation and restoration of cultural heritage. Proceedings of the 1995 LPC Congress. EPFL, Lausanne. 773 pp.

Krumbein W. E., A. A. Gorbushina, J. Valero, C. McCullagh, M. Young, P. Robertson, M. Vendrell, H.-L. Alakomi, M. Saarela, I. Fortune, L. Koscziewicz-Fleming (2006). Investigations into the control of biofilm damage with relevance to built heritage (BIODAM). Historic Scotland, Edinburgh, 96 pp.

Krumbein W. E. and K. Jens (1981). Biogenic Rock Varnishes of the Negev Desert (Israel) an Ecological Study of Iron and Manganese Transformation by Cyanobacteria and Fungi. Oecologia 50(1): 25–38.

Kucera V. (2003). Changing pollution situation and its effect on material corrosion. p. 23–29 in Proceedings of the 5th EC Conference Cultural heritage research: a Pan-European Challenge, Krakow, May 2002.

Machill S., K. Althaus, W. E. Krumbein, W. E. Steger (1997). Identification of organic compounds extracted from black weathered surfaces of Saxonean sandstones, correlation with atmospheric input and rock inhabiting microflora. Organic Geochemistry 27: 79–97.

Owczarek M., M. Spadoni, A. De Marco and C. De Simone (1999). Lichens as indicators of air pollution in urban and rural sites of Rieti (Central Italy). Fresenius Environmental Bulletin 8 (5–6): 288–295.

Saiz-Jimenez C (1995). Microbial melanins in stone monuments. Science Total Environment, 167: 329–341.

Sterflinger K. and Prillinger H.-J. (2001). Molecular taxonomy and biodiversity of rock fungal communities in an urban environment (Vienna, Austria). Antonie van Leewenhoek 80(3–4): 275–286.

Urzi C., W. E. Krumbein, T. H. Warscheid (1992). On the Question of Biogenic Color Changes of Mediterranean Monuments, (Coating -Crust – Microstromatolite – Patina – Scialbatura – Skin – Rock Varnish), in II. Int. Symp. for the Conservation of Monuments in the Mediterranean Basin, Musée d'Histoire Naturelle, Geneve, p. 397–420, edited by D. Decrouez, J. Chamay and F. Zezza, p. 519.

Viles, H. and A. A. Gorbushina (2003). Soiling and microbial colonisation on urban roadside stone: a three year study in Oxford, England. Building and Environment 38: 1217–1224.

Sources of Additional Information

The Journal of Biodegradation Society:
http:www.elsevier.com/wps/find/journaldescription.cws_home/405899/description#description
Proceedings of seminar on biodeterioration
Art, Biology, and Conservation: Biodeterioration of Works of Art
Edited by Robert J. Koestler, Victoria H. Koestler, A. Elena Charola, and Fernando E. Nieto-Fernandez
Introduction to Biodeterioration
Introduction to Biodeterioration/Dennis Allsopp, Kenneth J. Seal, Christine C. Gaylarde.
2nd edition, Cambridge; New York: Cambridge University Press, 2004.
Biodeterioration of cultural property
Edited by Robert J. Koestler. London; New York: Elsevier Applied Science; New York, NY, USA 1991
Molecular biology and cultural heritage
Proceedings of the International Congress on Molecular Biology and Cultural Heritage, 4–7 March 2003, Sevilla, Spain/edited by C. Saiz-Jimenez. Lisse; Exton, Pa.: Balkema, 2003

Chapter 6
Stock at Risk

John Watt, Stefan Doytchinov, Roger-Alexandre Lefèvre, Anda Ionescu, Daniel de la Fuente, Kateřina Kreislová, and Augusto Screpanti

6.1 Overview

The damage functions discussed in Chapters 3 and 4 can be used to assess the impact of pollution by calculating the rate at which a given material will corrode or soil. If it is desirable for policy development to estimate a regional impact, there is one other vital data component required, which is an estimate of the amount of each material being affected that is known as the stock at risk. This vital area remains seriously under-researched, mainly due to the high cost and complexity of compiling inventories. To date actually, there are no stock-at-risk estimates of any heritage materials at a continental level. Some research groups and institutions have tried to develop maps of heritage at risk in different countries, frequently based on different criteria, choosing different parameters of risk and materials.

The exercise of estimating stock at risk can be undertaken at a number of different scales and this chapter looks at several:

- Continental
- National
- City
- District of city
- Monuments

A number of studies have sought to estimate the occurrence of materials in residential and industrial buildings but estimation of materials in heritage buildings is much more challenging. Material composition and stock of modern buildings can be modelled by estimating the amount of different materials in typical types of building and then estimating the numbers of each type in a given region. This relies on that fact that, within discrete regions areas and covering given timescales, buildings of a given type are frequently very similar to each other.

J. Watt (✉)
Centre for Decision Analysis and Risk Management, School of Health and Social Sciences, Middlesex University, The Burroughs, London NW4 4BT, UK
e-mail: j.watt@mdx.ac.uk

J. Watt et al. (eds.), *The Effects of Air Pollution on Cultural Heritage*,
DOI 10.1007/978-0-387-84893-8_6, © Springer Science+Business Media, LLC 2009

Heritage buildings are typically much more diverse since they may be survivors of earlier periods or are valuable since they are relatively rare and/or may be unique or unusual at the time of construction. A series of small scale studies are being undertaken to evaluate the potential of using identikits for some types of heritage or to estimate heritage building stock by inspection. Several, which represent different scales of inspection, are discussed here.

There are a number of potential approaches to estimating stock at risk on a wider scale which will be discussed. Economic estimates of repair costs can be added later to perform regional estimates of economic impact of exposure of heritage to air pollution damage and to make targeted risk maps.

6.2 Stock at Risk at Different Scales

It is clearly both possible and desirable to map the stock of materials at risk at a wide range of geographical scales but usually five levels are considered:

Continental	exemplified here by Europe
National	exemplified here by the Italy, Norway, Czech Republic and the USA.
City	exemplified here by Madrid, Milan
District of city	exemplified here by Paris, Venice
Monuments	exemplified here by S. M. della Vittoria and Aldobrandini Villa in Rome

These levels do not always correspond directly to definitions used about what constitutes heritage. For example, perhaps the best criteria for definition of cultural heritage are indicated in the UNESCO – Art. 1 of the *World Heritage Convention*, which defines the following components of cultural heritage, which may overlap some of the scales described above.

Monuments: architectural works, works of monumental sculpture and painting, elements or structures of an archaeological nature, cave dwellings and combination of features, which are of outstanding universal value from the point of view of history, art and science.

Groups of buildings: groups of separate or connected buildings which, because of their architecture, their homogeneity or their place in the landscape, are of outstanding universal value from the point of view of history, art and science.

Sites: works of man or combined works of nature and man, areas including archaeological sites which are of outstanding universal value from the historic, aesthetic, ethnological or anthropological point of view.

The following sections briefly describe projects at each scale and outline some of the methods that have been used to obtain and use data at these different scales. This section draws on the extensive work related to the assessment of the effect of weathering and pollution on cultural heritage buildings where a lot of research on the determination of the "stock at risk" has been carried out.

6.2.1 Continental or Regional

There have been a number of projects that have included assessments of the stock at risk at a regional scale. One recent example is the work related to the implementation of the 1996 Oslo Sulphur Dioxide protocol (ApSimon and Cowell, 1996). This project used estimates of residential floor space per capita in countries which form part of the United Nations Economic Commission for Europe (UNECE) and a spatial distribution of buildings based on published population density data. Estimates of the materials in each unit of floor-space were based on data for Prague and Stockholm. Combining these datasets allowed a first order estimate of the total stock of materials at risk in Europe and so estimates of the savings in repair and maintenance costs for different pollution reduction scenarios. However, as would be expected, mapping at this scale involves a great deal of generalisation, limited categorisation of materials, and no mapping of the location of individual buildings.

The European Cultural Heritage is one of the richest in the world and the UNESCO (United Nations Educational Scientific and Cultural Organization) list of heritage with universal importance can be used for mapping. 387 of the UNESCO list of 745 sites are situated in Europe. After separation of the national parks and reserves etc., 243 heritage sites have been placed on the European EMEP (50×50 km grid) projection map using exact geographical coordinates (very often missing in the official list) using the Arc View programme. UNESCO sites are very different to each other, ranging from a single monument to sites like "Rome and Vatican city", which according to the Italian conservation authority includes more than 2,900 monuments. For a better presentation of the concentration of heritage in a single UNESCO site, the sites have been divided into three categories. The sites of different categories are presented with different symbols on the European map (Fig. 6.1).

Small ball – single monument selected from the Cultural Heritage sites in Europe

Square – small group of monuments (2–3) situated in the same place selected from the Cultural Heritage sites in Europe

Triangle – large group of monuments selected from the Cultural Heritage sites in Europe situated in the same place (for example: old city of Rome and the Vatican city, old city of Florence etc.)

The distribution of the UNESCO monuments in Europe is not uniform, with a large part concentrated in some Mediterranean and Central European countries. For example, in Fig. 6.1

Sites indicated with a triangle: 40 of a total of 63 monuments are in the Mediterranean part and 14 in Italy;

Sites indicated with a square: 22 of 43 are in the Mediterranean part.

Sites indicated with a ball are quite uniformly distributed.

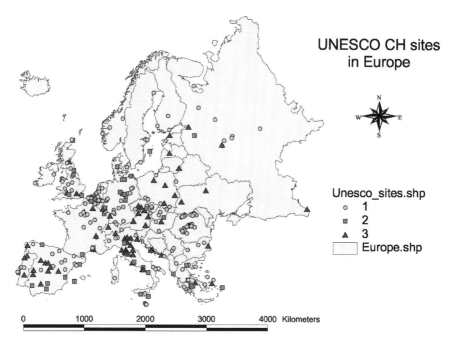

Fig. 6.1 UNESCO cultural heritage sites in Europe

The potential of such maps for the development of policy is demonstrated by the fact that a consequence of the higher concentrations of monuments in the Mediterranean region and in some Central European countries may be that more attention and resources for their preservation are needed in this part of Europe. This highlights the need to develop regional studies and valuations of effects of air pollution in these areas.

6.2.2 National

A range of methods exist for estimating the stock of buildings, including the use of large-scale ordnance survey maps, census type data, remote sensing and photogrammetry. Land-cover information does not provide data on the number or type of buildings, so the satellite data can be used to indicate the spatial distribution of the building stock, but an estimate of total stock in terms of buildings themselves is required. This can either be calculated "bottom-up", attempting to count individual buildings from maps, or "top-down" using census data. For commercial and residential buildings, the "top-down", census information on the number of buildings is available from separate sources, according to the types of buildings of interest. When multiplied by "identikits" (generic building types that are developed to represent the dominant styles of building found within a region and to provide estimates of the average proportion of different materials used in their construction) the census data can be used to estimate the national stock at risk. Total stock data which is available for the UK include:

- the number of dwellings (by type);
- the stock of industrial and commercial floor space;
- the number of schools (by type); and
- the stock of selected historic building

Because of the large variation in building type, this method is not applicable for heritage, although there may be potential in using national registers of heritage where the information is available in digital form.

Alternatively, studies have been undertaken to evaluate the potential of using published information from sources such as good quality tourist guides. The different tourist guides reports more or less detailed information. The Italian Touring Club Guide (ITCG) was chosen for one such study, from which information on the objects with historic and cultural value was extracted. As an example French heritage

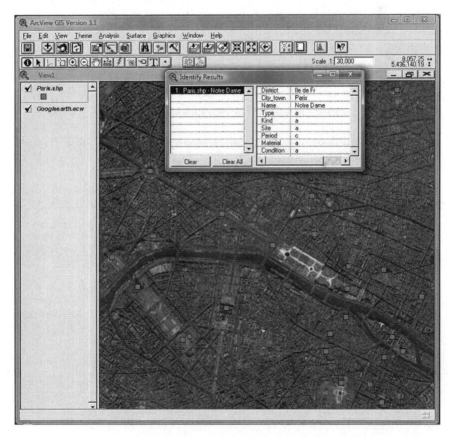

Fig. 6.2 Illustration of GIS-based stock-at-risk map for objects selected from ITCG Paris sites. After "clicking" on the single point of a monument it is possible to "enter" into the chosen city data set showing all available information regarding the single cultural object

objects were selected from the ITCG (actually for France there are 3 ITC Guides: for Paris, for the big cities and for the small towns) in the following ways:

level 1 – Paris: there are some 23 monuments indicated. Location of objects may be entered into a GIS data base and shown on a city map (Fig. 6.2), over which it is possible to overlap the yearly urban air pollutant concentration maps at high resolution (for example 2×2 km). So from the map it will be possible to see the real air pollution situation to which the different monuments are exposed and to calculate the dose-response corrosion function for the single materials. This type of risk map is a useful preliminary indicator for development of policy.

level 2 – big French cities: In France there are 19 major cities, such as Marseilles or Toulon. In some, as examples or special cases, the city maps with the heritage objects have been produced.

level 3 – relatively small Cultural Heritage towns: The study extracted some 18 small French towns such as Avignon and Chartres from the ITCG. In this case, it was considered sufficient to simply extract the number of monuments and map them on the national map as numbers, without the need to develop a town map, since the air pollutant situation is not complicated, except for some cases where there are important industrial areas.

Such tourist map-based inventories are only necessary where computerised national inventories are not available. The basic available information that was extracted from the Guide allowed development of different types of maps at city, regional, country and continental levels using the Arc View GIS System. Behind the maps the system allows the introduction of a data set tied to each location and with a simple "click" on the point of the town it is possible to "enter" into the chosen town data sets showing all available information regarding the cultural objects of the town. In the case of big cities "clicking" on the city point opens a detailed city map in which the points indicate the spatial distribution of the monuments. "Clicking" on the single point of Monument makes it possible to "enter" into the chosen city data set showing all available information regarding the single cultural object. On the city map it is possible to overlap the air pollution concentrations maps, or corrosion or damage maps. The manual entry of the data to reflect desirable features for stock-at-risk mapping, permits selection of single monuments by type, construction period or type of material.

Figure 6.3 shows a simple map of France with the UNESCO sites, the cities and the distribution of some important French monuments. Such a map, of course, represents a huge over-simplification but serves as an illustration of a possible method of mapping heritage. Use of tourist guides may be defended in that they offer a focus on internationally, nationally and regionally famous monuments but this is prone to a huge selection bias and does not really take materials into account.

Examples of stock-at-risk mapping from three representative countries: Italy, the Czech Republic and Norway are presented to demonstrate some of the work that has been done and some of the potential that such maps have for risk management decision making. Italy, Norway and the Czech Republic were chosen to exemplify different climate, history and air pollution, which determine the risk for the respective cultural heritage.

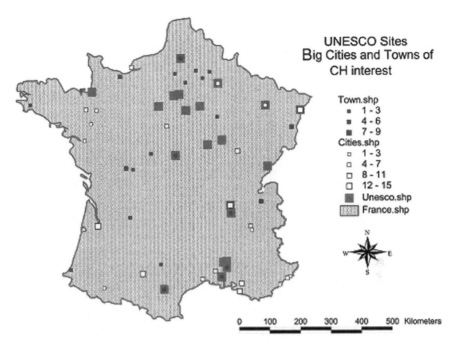

Fig. 6.3 French UNESCO cultural heritage sites and selected from ITCG French "big cities" and "towns". It can be seen that such maps only represent a tiny fraction of the heritage of a country but may focus on the most noteworthy or famous

6.2.2.1 Italy

A very detailed national scale survey has been carried out in Italy to develop a "risk map" of cultural heritage (The Risk Map of Cultural Heritage in Italy), which included around 50,000 single cultural monuments. The mapping process involved a series of thematic overlays relating to 40 different types of environmental danger for over 8000 municipalities. On the data base a description of every object (in an external environment) is included indicating the main materials used for its construction, which allowed estimation of the quantity of the main stone materials used.

Using the database it is possible to map the risks to individual buildings, for example the risks of static-structural movement (due to earthquakes, floods, landslides, etc.), relate to the distribution of slender towers, or the distribution of stone buildings in aggressive environments. The resources required to complete such a study are very large. As the inventory is done at a municipality level the individual areas on the map are the respective municipalities (Fig. 6.4 is an example). From the maps it can be seen that the heritage objects in Italy are distributed throughout the national territory, with areas of higher concentrations of heritage in the most important historical cities: Rome, Milan, Venice, and Naples.

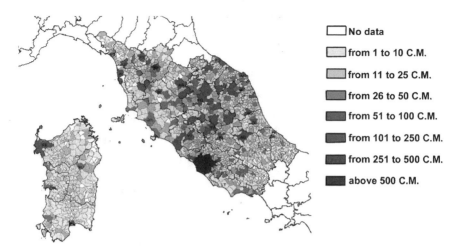

Fig. 6.4 Map of distribution of cultural monuments in central Italy

6.2.2.2 Norway

The cultural heritage archive in this country is based on the criteria that all buildings and artefacts constructed before 1615 are considered cultural monuments. Figure 6.5 shows the stone buildings (177), differentiated on a map by using ArcGis software. A general indication of the type of monument is included in the data base for every object.

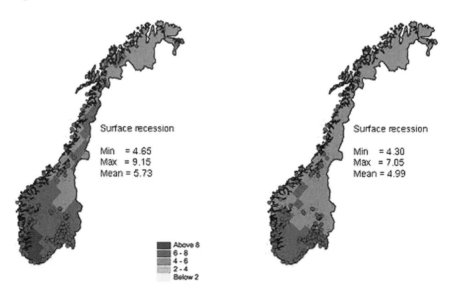

Fig. 6.5 Norwegian limestone corrosion map for the years 1990 (*left*) and 2000 (*right*), developed on the basis of the MULTI-ASSESS dose/response function, with Norwegian cultural heritage stone buildings (before 1615). EMEP unified model, 50×50 km grid. Unit is $\mu m\ yr^{-1}$

From the figure it can be seen that the monuments are not distributed uniformly throughout the territory, which is due to climate and the differing historical development of different parts of the country. This map only shows the stone cultural monuments, since dose-response functions exist for stone, but in Norway many cultural monuments are made of wood, for which there are no dose/response functions available.

6.2.2.3   Czech Republic

The Czech Republic has developed a good inventory of the national cultural heritage, organised by category. The stone monuments are divided into different types: castles, chateaus and ruins. All monuments are included on a map, using ArcGis software. A general indication of the type of monument for every object is included in the database. Figure 6.6 shows the spatial distribution of castles introduced on the corrosion maps for 1990 and 2000 developed on the basis of dose/response functions from the MULTI-ASSESS project. It is clear that these types of artefacts are quite uniformly distributed throughout the national territory; in some areas they are exposed to a harmful environment due to air pollution.

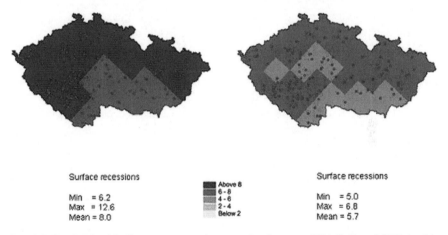

Surface recessions

Min   = 6.2
Max   = 12.6
Mean = 8.0

■ Above 8
■ 6 - 8
■ 4 - 6
■ 2 - 4
Below 2

Surface recessions

Min   = 5.0
Max   = 6.8
Mean = 5.7

Fig. 6.6 Czech Republic limestone corrosion map for the years 1990 (*left*) and 2000 (*right*), modelled on the basis of the MULTI-ASSESS dose/response function, with heritage castle sites. EMEP unified model, 50 ×50 km grid. Unit is μm yr^{-1}

6.2.3 City Scale

Many monuments are situated in cities where they are exposed to urban air pollutants, which means that there is a need for application of urban air pollution models to produce heritage risk maps.

Mapping at this scale allows a much higher resolution and, potentially, a much greater precision. A good example of work at this scale is described in Haagenrud (1997). In this study estimates of buildings and materials are based on Building

Registers, verified by inspection of individual buildings. The spatial distribution of buildings can be obtained from street maps. In cases where there are groups of similar buildings the "identikit" approach can be combined with this "high resolution" data, since there is a relatively uniform building type within these limited scales.

In the case of heritage this provides a very detailed knowledge of the location of individual buildings and, in countries such as Norway, detailed lists of the materials used. However, it still requires further investigation in cases where location of materials within the building and the condition of the materials is to be recorded.

Two examples of cases of cities with good cultural heritage inventories, Madrid and Milan, are presented.

6.2.3.1 Madrid

Here the local authority has developed an inventory of its heritage, dividing the objects into movable and immovable heritage. A total of 1,618 movable items such as sculptures, fountains etc. have been collected, along with 260 immovable items including buildings, bridges etc. Figure 6.7 presents both types of cultural heritages, which are both concentrated in the centre of the city.

Maps of spatial distribution of sulphur dioxide on Madrid City for the year 2000 and 2004, with grid cell size of 500×500 metres were developed using Kriging Interpolation. For the year 2000, sulphur dioxide is concentrated in the downtown area, which has high traffic density, and gives values near to $40 \, \mu g \, m^{-3}$. Such combined maps are useful for showing owners, managers and other stakeholders the extent of air pollution risk to their object (see also Fig. 6.8).

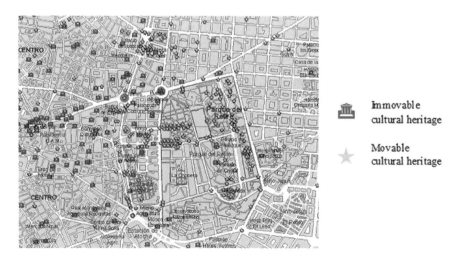

Immovable
cultural heritage

Movable
cultural heritage

Fig. 6.7 Detail of the movable and immovable cultural heritage distribution in Madrid

6.2.3.2 Milan

This study was run from ENEA and Instituto Centrale per il Restauro di Roma. The inventory of cultural heritage in Milan formed part of the Map of Risk of Italian Cultural Heritage discussed above. In all, 1,200 cultural monuments were mapped and Fig. 6.8 shows the distribution of the stone monuments. By using Kriging interpolation, the maps of spatial distribution of the concentration of sulphur dioxide on Milan for the year 2000 were developed. It can be seen from the map that a large part of the cultural heritage in Milan is exposed to relatively high concentration of SO_2 (the data are average annual concentration of the pollutant).

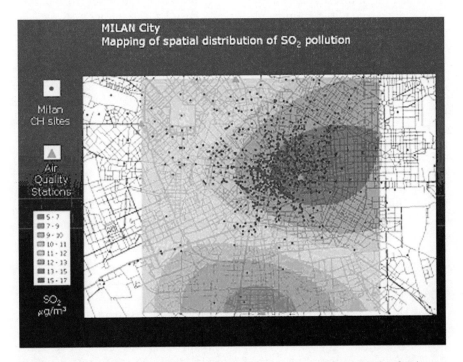

Fig. 6.8 Spatial distribution of stone cultural heritage in Milan with SO_2 pollution for the year 2000, modelled by the Kriging tool in GIS

6.2.4 District Level

Three case studies are presented below, using methodology developed by the Laboratoire Interuniversitaire des Systèmes Atmosphériques, Université Paris XII, which consisted of a field inventory façade by façade, building by building, monument after monument (see Case Study – Evaluation by a Direct Measurement Method of the Stock of Materials at Air Pollution Risk on the

Façades in three Cities inscribed on the UNESCO List: Paris, Venice and Rome – at the end of this chapter).

A *Direct Measurement Method* is not commonly used for the evaluation of the Stock of Materials exposed to the Air Pollution Risk because they require intensive work in the field measuring the surface occupied by each type of material once its nature determined (stone, mortar/rendering/plaster, painting, brick, metal, glass...). This very simple method permits an evaluation without any theoretical modelling or generalisation, simply by direct observation, counting and simple measurement. It was applied to representative parts of three cities inscribed on the UNESCO World Heritage List: *Paris, Venice* and *Rome*. The comparison of the results obtained demonstrates quantitatively that the nature and the quantity of materials used for the construction of the buildings and of the monuments in a city are intrinsic characteristics of the city, resulting from its geological environment (sources of natural materials), its history and its cultural, architectural and handicraft traditions, but also from the methods and materials used for maintenance and for restoration works.

In Paris, for example, the study was based on a map named "Height of floors" at the scale of 1: 2000 and showed that the nature of the materials employed in construction was Lutetian Parisian limestone, rendering/mortar/plaster, painting, brick, metal and modern glass.

6.2.5 *Individual Buildings*

In cases where individual buildings are of interest it is possible to combine a much greater range of high-resolution data. The data can be obtained from a range of spatially and non-spatially referenced sources including detailed photographs. Mapping can include materials and decay features and if suitable photographs are available, changes in decay with time.

Different studies regarding the maintenance, conservation and restoration costs of single monuments have been developed. The methodology used is demonstrated on two objects: the Aldobrandini Village House, located in the Roman province of Frascati and the Santa Maria della Vittoria Church located in the town centre of Rome.

6.2.5.1 Aldobrandini Village House

The Google Earth Programme was used for the determination of the exact location of the villa (This programme not only provides the coordinates of the country, city, area or even the street or monument concerned, but it also shows satellite pictures of it, Fig. 6.9).

After establishing the location of the villa, the total surface of the frontal, lateral and backside facades of the principal building was calculated. The length of the frontal facade measures 52 m, while its height and width is 14 m. From

Fig. 6.9 Aldobrandini Villa (Frascatti, Rome). Image derived from the satellite map of Italian Ministry of Environment, Territory and Sea

these measurements it was estimated that the front facade surface area is 728 m^2 (without the upper tower), and the left and right lateral facades areas are 196 m^2 each. The rear facade is different from the front, since it has a protruding donjon of 15 m with a thickness of 5.50 m, Fig. 6.10.

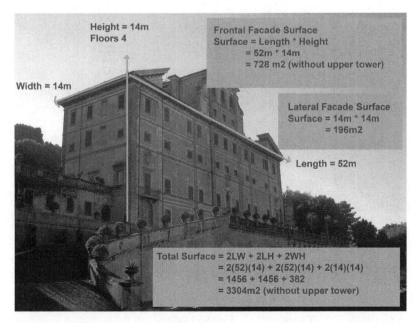

Fig. 6.10 Villa Aldobrandini with façade areas

Fig. 6.11 Villa Aldobrandini

The building has a sloping roof with tiles and iron zinc plated galvanized eaves, with a stucco cornice. The windows have wooden shutters and iron gratings. The facades prime material is plaster with mortar lime and "pozzuolana" in the background, finished with a thin layer of lime stucco and marble powder, and coloured with a water diluted varnish, Fig. 6.11.

The Villa is decorated with different types of windows, which were classified and measured, in order to estimate the total glass surface of the facades. They range from small square/oval windows, to medium and large rectangular ones, to arched ones of different sizes. The building includes also a full-size balcony in the rear facade in the last floor, facing the backyard of the Villa, Fig. 6.12.

Using this data the total amount of glass in the Villa's main building is estimated to approximately 200 m^2. If the frontal facade surface is 728 m^2 it means that this side of the building has 39% glass of the total surface, the

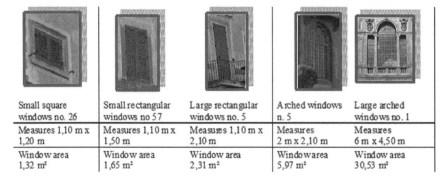

Small square windows no. 26	Small rectangular windows no 57	Large rectangular windows no. 5	Arched windows n. 5	Large arched windows no. 1
Measures 1,10 m x 1,20 m	Measures 1,10 m x 1,50 m	Measures 1,10 m x 2,10 m	Measures 2 m x 2,10 m	Measures 6 m x 4,50 m
Window area 1,32 m^2	Window area 1,65 m^2	Window area 2,31 m^2	Window area 5,97 m^2	Window area 30,53 m^2

Fig. 6.12 Different types of typical Roman windows in the Villa

backside facade has 43% and the lateral facades 9% each one. If the frontal facade has 77.9 m^2 of glass, and, after subtraction of the entrance space (8 m^2 = 2 m length × 4 m high) it can be assumed that the remaining 642 m^2 is plaster.

6.2.5.2 Santa Maria della Vittoria Church

As for the Frascati Villa, the Google Earth Programme was also used to find the coordinates of the Santa Maria della Vittoria Church, located in the town centre of Rome. This Roman baroque church dates from 1620, has a plain solemn travertine marble facade, and an interior, rich with important sculptures, Fig. 6.13. The church has a single wide nave under a low segmental vault with three interconnecting side chapels behind arches. The masterpiece of the chapel is the "Ecstasy of Santa Theresa" by Bernini located in the left of the altar.

Since the frontal facade of the church has a complex outline, it was divided into different geometrical figures, Fig. 6.14, each of whose area was calculated and summed. These measurements were taken from a digital prospect of the building, to understand the different spaces of the antique construction.

Fig. 6.13 Views of the frontal and side facades of the Santa Maria della Vittoria church in Rome

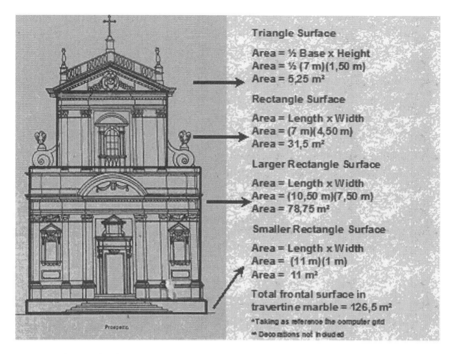

Fig. 6.14 Areas of different parts of the frontal façade of the Santa Maria della Vittoria church in Rome

6.2.6 Deciding What Information Is Needed for Mapping Stock at risk

It is clear that estimating stock at risk for such diverse monuments and objects is challenging. National resisters exist in most countries to provide a catalogue of the national heritage but these do not use a standard methodology for collation, for selection criteria or, most frustratingly in this context, for recording composition (materials). Studies need to develop specific strategies for sampling and inspection, in order estimate material at risk. If a European scale estimate is to be made, for instance, some standard criteria will need to be agreed.

It is possible to collect a large amount of information but it may not all be that necessary or appropriate and so there must always be a balancing of needs against the availability of data. This is important to ensure that only appropriate data is collected and processed as it is all too easy to collect a lot of data that will add nothing to the determination. The appropriateness of the data will also be related to the scale of the survey and the desired end point. In the case of the Strategic Stone Study in the UK, it was considered more important to collect a wide-scale and representative dataset rather than a large amount of detail at a limited number of buildings which may give a distorted outcome.

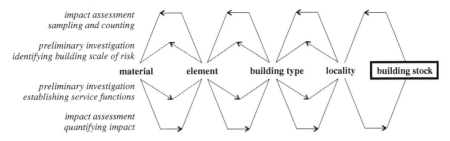

*impact assessment
sampling and counting*

*preliminary investigation
identifying building scale of risk*

material   element   building type   locality   building stock

*preliminary investigation
establishing service functions*

*impact assessment
quantifying impact*

Fig. 6.15 Illustration of the "top down" and "bottom up" approach to collecting data.

The key areas to consider are:

- Spatial distribution of buildings
- Materials inventory
- Condition survey

The resolution and detail for each dataset needs to be comparable and some form of sensitivity analysis can be used to determine the "weakest link in the chain". For example, there is little point in having the location of every building if there are no records on the origin of the stones used in those buildings.

Also when considering deterioration and weathering there are two approaches – "top down" or a "bottom up". In the former the emphasis is on the building stock and then looking at detail. The latter begins with the material or component and looks how the stock will be affected. These two approaches are illustrated in Fig. 6.15, which is taken from a project looking at the effects of long-term climate change on buildings (Fedeski and Johns, 2001).

In the Strategic Stone Study these extremes could be represented by a survey of all examples of one stone type and a full survey of all materials in a limited number of buildings. So the key message is to use the correct data resolution for the desired scale of mapping – and make sure that the desired information will be collected.

6.2.6.1 Data Sources

As an example from the UK, a variety of sources are available to assist in trying to obtain a good background of information prior to going out on site. This includes:

- National Census online for the details of the number of properties in a parish
- Directory of listed buildings from English Heritage
- On-line photographic collections (including Images of England, Viewfinder and The Frances Frith Collection)
- British Geological Survey (BGS) maps of known quarries
- Publications – relating to specific stones, buildings or localities
- Archive documents

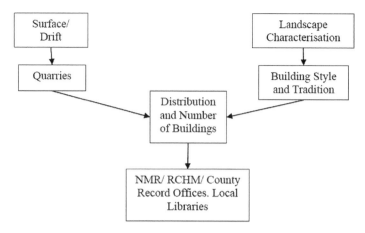

Fig. 6.16 Two routes that can be followed in order to identify relevant buildings that can be included in a sample of buildings for a parish or region

The most important use for these sources is to identify localities and individual buildings that can be included in the sample of buildings for a parish or region. There are two routes that can be followed in order to identify relevant buildings and these are illustrated in Fig. 6.16.

The left-hand route starts with the surface geology and attempts to identify areas where stones suitable for building are likely to be found. If this information is then combined with the distribution of known current or historic quarries (data held by the British Geological Survey) then the likely location of parishes/regions with significant numbers of vernacular buildings and structures in stone can be plotted.

The right-hand route takes a more architectural approach and has as its starting point the characterisation of landscapes in terms of geology, geomorphology, and natural environment. This provides a basis for general statements to be made about the landscape including the types of buildings and materials used.

Combining these two routes allows more detailed archive work and surveys to be concentrated in the most promising locations – that is those areas where the traditional vernacular architecture is based on local natural stone. There will, of course, be individual buildings, such as country houses and churches that were of sufficient status to warrant the importation of stone from elsewhere. This will be true for all areas – both those where stone is widely available and those where it is not part of the local tradition. However, these buildings are likely to be of sufficient status that they can be readily identified through publications and archives.

6.2.6.2 The United States

There are two mandates in the United States to identify buildings with historic value, the National Historic Preservation Act of 1966 (and amendments) and President Nixon's Executive Order 11592, which mandated that all states be surveyed to "locate, identify and register all historic properties". The universe of historic buildings is determined by the State Historic Preservation Offices (SHPO), whose decisions are recorded in the National Register of Historic Places, maintained by the National Park Service. Because the National Register of Historic Places initially excluded structures that are commemorative in nature, other data sources were developed to inventory monuments and sculptural resources. A national survey of public art was performed in the mid-1980s, under the auspices of the Save Our Sculpture initiative. The National Park Service also maintains an inventory of monuments located on national battlefields and in Washington, DC.

Building materials sensitive to pollution (such as masonry, bronze & copper) can be located by reviewing National Register of Historic Places documentation, followed by a field verification to confirm masonry types. In some cases, building descriptions do not include materials, particularly for historic district documentation prepared prior to 1987, when guidance was published on how to complete the national register form. Thus field surveys were needed, which attempted to identify as many occurrences of sensitive material as possible. The field survey teams included a member trained in geology and all team members were schooled in identifying specific metal and stone types.

6.2.7 Indicators

As discussed above "identikits" have been used in the past to estimate the stock of different materials at risk in different regions. The size and composition of the external building envelopes provides an estimate of the area (m^2) of brickwork, concrete, etc in a single building, in each region, for each type of building.

The study combined the "bottom-up" and the "top-down" approaches (Yates, 1998). Building identikits were devised and tested which represented a typical description and specification of the external fabric of selected types of building (e.g. detached house, office block or semi-detached house). The identikit type indicated the surface area of building material for the type of building. Table 6.1 below is an example of a building identikit.

The identikit represents a weighted average of different architectural styles of semi-detached buildings, in an area. It shows, for example, three different types of roofing material, which would never be present on a single building.

It would be highly desirable to extend this to heritage so that maps prepared on the basis of the historical period or kind of the monument (churches and convents) could be used as "indicators" of cultural heritage, which are largely distributed in Europe.

6.2.7.1 Gothic Cathedrals and Convents

A preliminary pilot evaluation of this method is shown in Fig. 6.17. The Medieval stone Gothic Cathedrals and Convents of France were extracted

Table 6.1 Example of a building identikit

	Semi-Detached House
Flat roof area (m^2)	12.8
Pitched roof area (m^2)	54
Net external wall area (m^2)	84
Window frames, external doors and door frame (m^2)	9
Roof drainage (m^2)	3
Total external envelope (m^2)	165

Predominant Materials/Elements	Model (%)	Quantity (m^2/Building)
Flat roofs		
Felt (flat roofs on garages)	50	6.4
Asbestos cement (on garages)	50	6.4
Pitched roofs		
Clay tiles	25	13.5
Slate tiles	50	27.0
Concrete tiles	25	13.5
External walls		
Brickwork (bare)	58	48.7
Brickwork (painted)	10	8.4
Render	30	25.2
Timber framing or cladding	2	1.7
Windows and doors joinery	70	6.3
Aluminium/Metal	30	2.7
Roof drainage		
Plastic (gutters/pipes)	71	2.1
Cast iron (gutters/pips)	29	0.9

from the tourist guide data sets described earlier. This permitted the preparation of maps only for the "indicators". Such a use of indicators for assessing amount of material at risk (and therefore estimating damage costs via the economic assessments discussed in Chapter 7) have potential only in a situation where the identikit developed represents a realistic estimate of the average amount of material utilised per unit area, as is the case for identikits developed for suburban housing for example. Land use maps and remote sensing may then be utilised to estimate the total material at risk. The situation is far more complex when cultural heritage is considered, since buildings are much more individual in scale and design, even when built in the same period. Materials used may reflect local sources on some occasions and not on others. There may be value in this approach but it is almost certain that individual site evaluation would be needed to audit the material in each cathedral – which given their size and importance may well be a worthwhile enterprise.

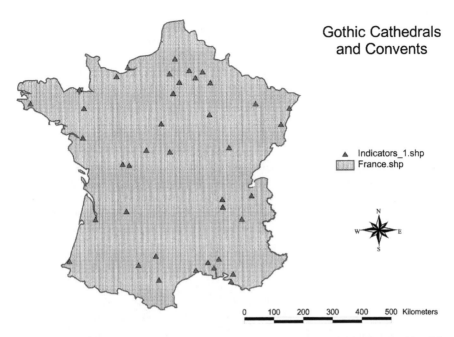

Fig. 6.17 Spatial distribution of the Gothic cathedrals and convents in France: identikit elaborated on the basis of information extracted from ITCG. Use of such indicators has potential where material use is relatively uniform. This is not the case for this type of building, since each one is unique

6.2.7.2 City of London Churches

Amongst the huge office blocks of the City of London, which is the heart of the London's financial quarter and covers only one square mile, are hidden around 50 current or former churches and other places of worship, either whole, changed into offices, or in ruins. Once there were nearly 100 parish churches within the City boundaries but the Great Fire of London, the migration of residents to the suburbs, and World War II bombs have reduced that figure. After the Great Fire in 1666 only 21 of the 108 churches remained intact. By 1670 it was agreed that 51 should be rebuilt with commission going to Christopher Wren who was assisted by his colleague and fellow scientist Robert Hooke. Today 23 Wren churches survive after Victorian rationalisation and the Second World War (Hatts and Middleton, 2003). This history and the restricted area made the City of London an excellent choice of location to test the development of identikits for historic churches (Andrews, 2006). A total of 46 churches were included in the study.

During the study, a personal visit to each of the churches was made. An inventory of the material make-up and dimensions of the church was made using a specially developed pro forma as a guide. The lengths, breadths and heights of the churches were estimated by direct examination of the buildings (by counting the number of blocks of stone or bricks in a row, making the width and counting the number of blocks in a column making the height and multi-plying the results from the count to the measured size of the block or brick). Having height and length, the surface was easily deduced. This entire surface was attributed to the identified constituting materials according to their pro-portions. The surface of the apertures (windows, doors) was determined by estimating their area.

An average size and amount of construction material which comprised the external envelope of the churches can be used to develop an identikit for churches. The churches have been grouped into stone churches and brick churches and the possibility of drawing up an identikit evaluated separately.

Out of the 46 churches, 24 of the churches were constructed mainly from stone. Five of the churches were constructed from stone and brick, with the predominant material being stone. Also 17 of the churches were constructed from a combination of stone and brick, of which, nine were predominantly brick with more than 50% of the total visible surface area composed of brick.

Brick Churches

Table 6.2 represents the quantities of brick, stone and glass in the brick churches. The weighted average of the quantities of materials in the brick churches has been estimated.

Table 6.2 Quantity of stone, brick and glass in the brick churches

Brick churches	Quantity of brick/m^2	Quantity of stone/m^2	Quantity of glass/m^2
St Benet Welsh Church	657.12	87.67	121.59
St Botolph without Bishopsgate	554.24	194.14	96.21
St Botolph without Aldersgate	709.92	195.60	77.07
All Hallows on the Wall	554.67	149.66	9.99
St Andrew by the Wardrobe	660.00	96.90	80.09
St Anne and St Agnes	502.00	39.05	75.55
St Helen Bishopsgate	310.05	290.64	65.12
St James Garlickhythe	327.94	296.61	94.15
St Mary Abchurch	821.46	96.85	116.04
St Mary Le Bow	651.57	543.36	77.82
St Mary At Hill	442.55	191.27	30.10
Weighted Avg. of materials	562.90	198.30	76.70

Identikit for Brick Churches

Thus, the identikit (Table 6.3) produced from this study excludes any material investigation of the roofing of the churches. Five types of roofing design were observed about the churches from literature. They are flat roof, coved roof, barrel-vaulted roof, groined roof and domed roof. Some churches have a combination of roofing designs.

Table 6.3 An identikit for brick churches

	Brick Churches
Flat roof area (m^2)	–
Coved roof (m^2)	–
Barrel-vaulted roof (m^2)	–
Groined roof (m^2)	–
Domed roof (m^2)	–
Total external envelope (m^2)	837.90
Predominant materials/Elements	**Quantity (m^2/Church)**
Flat roof Material not investigated	–
Coved roof Material not investigated	–
Barrel-vaulted roof Material not investigated	–
Groined roof Material not investigated	–
Domed roof Material not investigated	–
External walls	
Brickwork	562.90
Stonework	198.30
Windows Glass	76.70

Stone Churches

The dimensions of the stone churches were estimated in the same way as for the brick churches. The weighted average of these materials has been calculated. Table 6.4 shows the quantity of materials in the external envelope of the stone churches

Table 6.4 Quantity of stone, brick and glass in stone churches

Stone churches	Quantity of stone/m^2	Quantity of glass/m^2	Quantity of brick/m^2
St Olave, Hart St	275.44	51.83	
St Peter Upon Cornhill	678.60	22.03	175.46
St Sepulchre	484.32	112.61	
St Stephen Walbrook	920.59	24.13	
St Vedast, Foster Lane	517.02	54.28	83.63
City Temple	129.32	100.47	
The Dutch church	640.55	42.39	
St Bartholomew the less	344.81	66.36	240.84
St Botolph without Aldgate	637.55	46.02	237.21
St Bride	962.73	78.02	
St Clement	563.29	18.41	
All hallows by the tower	431.13	212.94	169.18
St Edmund the King and martyr	825.17	67.36	
St Giles Cripplegate	1,082.31	110.75	

Table 6.4 (continued)

Stone churches	Quantity of stone/m^2	Quantity of glass/m^2	Quantity of brick/m^2
St Katherine Cree	474.40	42.79	
St Lawrence Jewry	867.69	213.06	
St Magnus the Martyr	585.30	53.38	
St Margaret Lothbury	311.02	37.60	
St Margaret Pattens	1,146.90	83.52	
St Margaret Ludgate	230.84	24.02	
St Mary Aldermary	144.20	40.08	
St Mary Woolnoth	908.72	22.97	
St Michael Paternoster Royal	801.34	65.22	176.32
St Nicholas Cole Abbey	833.49	120.26	
St Michael Cornhill	934.72	90.68	
St Etheldreda's Chapel	223.62	36.51	
St Mary Moorfields	153.97	8.88	
St Ethelburga Centre	246.60	22.35	
Weighted Avg. of Materials /m^2	584.10	52.61	38.66

Identikit for Stone Churches

Table 6.5 shows an identikit for stone churches developed from the table above.

Table 6.5 An identikit for stone churches

	Stone Churches
Flat roof area (m^2)	
Coved roof (m^2)	
Barrel-vaulted roof (m^2)	
Groined roof (m^2)	
Domed roof (m^2)	
Total external envelope (m^2)	675.37
Predominant materials/Elements	**Quantity (m^2/Church)**
Flat roof Material not investigated	–
Coved roof Material not investigated	–
Barrel-vaulted roof Material not investigated	–
Groined roof Material not investigated	–
Domed roof Material not investigated	–
External walls	
Brickwork	38.66
Stonework	584.10
Windows Glass	52.61

Estimating Total Stock at Risk of Materials From an Identikit

From the identikit tables above, it would theoretically be possible to estimate the total stock at risk of materials of all brick or stone churches once they have been mapped. This can be done by multiplying the number of brick churches by the average area in m^2 of the materials indicated in the above identikit. In practice, however, the tables reveal how unreliable this would be. Even in this deliberately limited case study, with its restricted spatial scale and its truncated history, there is a tremendous amount of variability. In this instance this is due to the fact that City Churches were seldom isolated, often one or more sides being built against, so that opportunity for exterior design was limited and hence Christopher Wren's tendency to restrict his ornament to the steeples. His plans were frequently irregular in shape as, for economy sake, he often utilised existing foundations (Cobb, 1962). This type of variation among a collection of heritage buildings is hardly unusual, however. Significant buildings such as churches, castles and major houses, were often built to individual plans and incorporating unique design features as befitting their importance to those commissioning them. This means that the use of identikits in this context is difficult.

6.3 Case Study – Evaluation by a Direct Measurement Method of the Stock of Materials at Air Pollution Risk on the Façades in Three Cities Inscribed on the UNESCO List: Paris, Venice and Rome

6.3.1 The Stock of Materials at Risk on the Façades in the Centre of Paris

The banks of the Seine have been included on the UNESCO List of the World Cultural Heritage (Fig. 6.18) since 1991 because they are studded with a succession of masterpieces including Notre Dame Cathedral and Sainte Chapelle, Louvre, Palais de l'Institut, Invalides, Place de la Concorde, Ecole Militaire, La Monaie, Grand and Petit Palais des Champs Elysées, the Eiffel Tower, Palais de Chaillot and Trocadéro gardens. Notre Dame and Sainte Chapelle were definite references in the spread of Gothic construction. The eastern and central parts of Paris, the Marais and the Ile Saint Louis have coherent architectural ensembles of Parisian constructions of the 17th and 18th centuries, while Haussmannian urbanism marks the western part of the city.

The present study consists in the evaluation of the stock of materials at risk of degradation (surface loss, soiling) due to atmospheric pollution, between the Sully Bridge on the eastern side, and the Pont-Neuf on the western side, including thus, the Ile Saint Louis, the Ile de la Cité and the right bank of the Seine facing these two islands (Fig. 6.18). This sector comprises the very centre of Paris but not the totality of the territory inscribed on the UNESCO List,

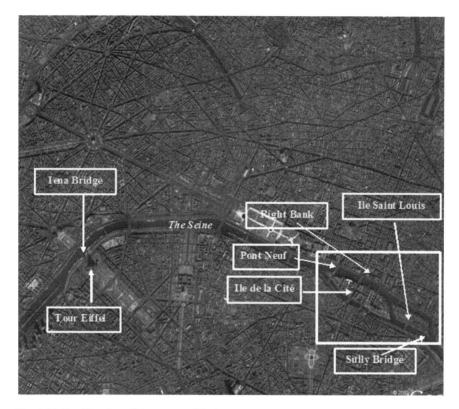

Fig. 6.18 Satellite view of the centre of Paris

which extends towards West as far as the Eiffel Tower; this study thus, represents roughly one-third of this territory and contains buildings dating from the 17th and 18th centuries, Haussmannian buildings (end of 19th and beginning of 20th centuries), as well as important monuments like Notre Dame and Sainte Chapelle dating from the Middle Ages.

The methodology consisted of a real in-the-field inventory, façade by façade, building by building, and monument by monument, based on the Paris Map at the scale of 1:2 000. The nature of the materials employed was determined by direct examination of the building façade (Lutetian Parisian limestone, rendering/mortar/plaster, painting, brick, metal, modern glass) and their proportions were roughly evaluated in percentage. The height of each building was estimated by counting the number of floors and attributing them an individual average height of 3 m. A control of this arbitrary height of 3 m per floor was performed using a laser beam measurement: the error does not exceed ±10%. The determination of the length of the façades was obtained by measurement on the Paris Map. Having height and length, the surface was easily deduced. This entire surface was attributed to the constituting materials according to their proportions. The surface of the apertures (windows, doors), classically considered by the architects equal to half of the total surface of the façade, was not deducted because it compensates for the roughness of the façade (sculptures, decoration, balconies...). In summary, the total calculated surface was attributed to the constituting materials and half of this surface arbitrarily attributed to the modern glass of the windows. Only the street facing, external façades were taken into account due to their direct exposure to pollution from traffic and the inaccessibility of interior private courts.

The real surface of historical monuments is theoretically available from the Architect in Chief in charge of the different monuments, but it was impossible to obtain this information for reasons of confidentiality. Thus, it was decided to measure these surfaces directly in the field according to the same methodology employed for the private buildings. In the Ile de la Cité, the quantity of historical monuments and official buildings is very high: Notre Dame Cathedral (Fig. 6.19), Hôtel Dieu Hospital, Paris Police Headquarter (Préfecture de Police), Commercial Court (Tribunal de Commerce), Law Court (Palais de Justice) and Sainte Chapelle. On the right bank of the Seine, three important monuments exist: the Town Hall (Hôtel de Ville de Paris) and two theatres: the Théâtre de la Ville and the Théâtre du Châtelet. There are only two historical monuments in the Ile Saint Louis: the Church Saint Louis-en-l'Ile and the Hôtel de Lauzun both belonging to the Paris City Patrimony. All the other historical buildings in this island are private property. Quays and bridges were not taken into account in this evaluation.

Fig. 6.19 The eastern part of the Ile de la Cité with the Notre Dame Cathedral (100% limestone) in the background. On the first plan, the Quai aux Fleurs with, on the left, behind trees of the Ile Saint Louis, an Haussmannian building (100% limestone) and some houses dating from the 17th–18th centuries partially in limestone, partially painted. View taken from the Hôtel de Ville

In total, the measurement of the length of each construction on the map of Paris, the counting in the field of the number of floors and the characterisation and evaluation of the respective proportions of the constituting materials were performed on the façades of 525 buildings and monuments in the centre of Paris giving an excellent statistical value to the results presented below.

The total surface in m^2 of the 525 façades of buildings and monuments of the Ile Saint Louis, Ile de la Cité and of the right bank of the Seine facing them, and the distribution of the different materials in these façades are given in Table 6.6, on the basis of a 3 m mean height. The surface of modern glass is arbitrarily estimated as half of the total surface of the façade.

The two historical monuments of the Ile Saint Louis have their façade entirely in limestone, covering $768\,m^2$, and the other six, in the Ile de la Cité accounting for $71,586\,m^2$.

These results demonstrate that the main material present in the façades in the centre of Paris is the Lutetian limestone (roughly 76%), followed by painting (15%) and then by rendering (7%). Brick (1%) and metal (0.02%) play a very minor role. These materials are displayed according to the architectural style dominating in each location. Thus, limestone dominates in the Ile de la Cité and on the right bank facing it, due to the presence of many important monuments and Haussmannian buildings, while painting and rendering are more important in the Ile Saint Louis, where the buildings dating from the 17th and 18th centuries have been recently restored and widely painted. Meanwhile, some "Noble Hôtels particuliers" dating from these centuries are in limestone. In the Ile de la Cité or on the right bank of the Seine some remarkable buildings are an assemblage of limestone and brick (Fig. 6.20). Constructions entirely made of brick are rare.

The geographical distribution, on a grid of 100 m × 100 m cells, of the total surface of façades, of the surface in limestone and of the percentage of limestone in the façades is given in Figs. 6.21, 6.22, 6.23. These confirm in more detail that limestone is in the majority in the western part of the studied area, meaning that the most important monuments and the highest number of Haussmannian buildings are concentrated in the Ile de la Cité and on the right bank of the Seine facing it.

The main risk for buildings and monuments in the centre of Paris is air pollution due to the traffic causing the soiling of façades by deposition of black carbonaceous particles, especially in the parts sheltered from rain, and the surface recession of these façades by erosion-dissolution in the parts exposed to the rain. The floods of the River Seine affect only the lowest level of the banks and only very exceptionally reach the streets, buildings and monuments (one or two times per century). Capillary rising of water loaded with salts at the base of walls are not of major importance in the centre of Paris thanks to the entire covering of the ground by impermeable pavement.

Table 6.6 Comparison of the percentages of different materials present in the façades of buildings and monuments in the three studied areas: The Centre of Paris, the Sestiere of Dorsoduro in Venice and the Via del Babuino in Rome

m²	Façades total surface	Façades total length	Limestone travertine Marble	Render/mortar/plaster	Painting	Brick	Metal	Modern Glass (estimated)
Total Paris Centre (Monuments)	525 façades = 200,305 m² (72,354 m²)	11,203 m –	15,933 m² = 76% (100%)	14,386 m² = 7%	29,907 m² = 15%	2,534 m² = 1%	420 m² =0,02%	(50% = 100,152 m²) –
Total Venice Dorsoduro	279 façades = 48,361 m²	5,241 m	4,808 m² = 10%	36,846 m² = 76%	–	6,676 m² = 14%	–	(50% = 24,180 m²)
(Palaces on Canal Grande)	(1995 m²)	–	(43%)	(51%)	–	–	–	(Glassy mosaics = 6%)
Total Via del Babuino (Rome)	70 façades = 16,913 m²	921 m	3,411 m² = 20%	902 m² = 5.4%	11,479 m² = 68%	739 m² = 4.4%	–	(50% = 8,456 m²)

Fig. 6.20 Western extremity of the Ile de la Cité: on the left, a building in limestone and bricks on the Place du Pont Neuf, in the centre the statue of Henri IV in bronze. In the middle distance, the Coupole of the Institut de France and, in the background, the Tour Eiffel are located on the left bank of the Seine

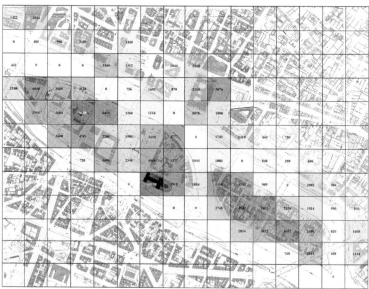

Fig. 6.21 Geographical distribution of the total surface (m²) of the façades of buildings and monuments of the Ile de la Cité, the Ile Saint Louis and of the right bank of the Seine facing them, on a grid with cells of 100 m × 100 m (floor mean height: 3 m)

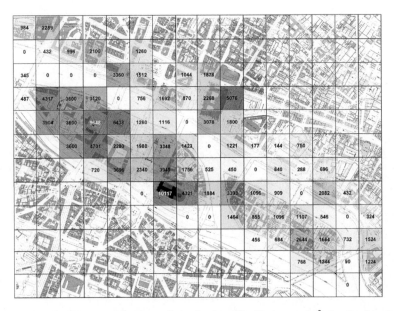

Fig. 6.22 Geographical distribution of the Surfaces in limestone (m^2) in the façades of buildings and monuments of the Ile de la Cité, the Ile Saint Louis, and of the right bank of the Seine facing them, on a grid with cells of 100 m × 100 m (floor mean height: 3 m)

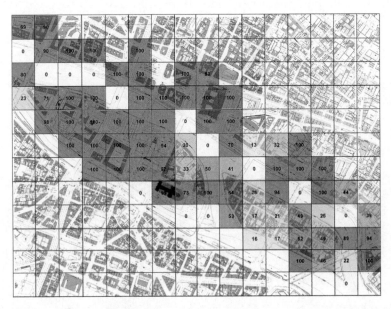

Fig. 6.23 Geographical distribution of the percentage of limestone in the façades of buildings and monuments of the Ile de la Cité, the Ile Saint Louis, and of the right bank of the Seine facing them, on a grid with cells of 100 m × 100 m (floor mean height: 3 m)

6.3.2 The Stock of Materials at Risk on the Façades in the Sestiere of Dorsoduro in Venice

Venice is inscribed on the UNESCO List of Mankind Cultural Heritage since 1987 for its unique artistic achievement, its considerable influence on the development of architecture and monumental arts and its incomparable series of architectural ensembles illustrating the age of its splendour. From monumental complexes such as Piazza San Marco (Basilica, Palazzo Ducale, Campanile, Procuratie) to the more modest residences of "calle", "calleselle" (narrow streets) and "campi" (small squares), including hospitals and charitable or co-operative institutions ("Scuole"), Venice presents a complete typology of mediaeval architecture with a relatively high homogeneity in the employed materials as it will be demonstrated.

For the evaluation of the Stock of Materials at Risk in Venice, a typical and representative area was chosen in Dorsoduro, one of the six administrative districts ("Sestriere") of Venice (Fig. 6.24). Dorsoduro comprises some important monuments (palaces, churches) representative of the Venetian architecture and a quarter with typical Venetian private houses.

The area where the inventory of materials was performed is located in the southern part of the City between the Canal Grande on the North, the Giudecca Canal on the South, the Rio Terra Antonio Foscarini on the West and the line joining the Palazzo Venier dei Leoni (Guggenheim Foundation) to the Spirito Santo Church on the eastern side. It represents a territory covering a

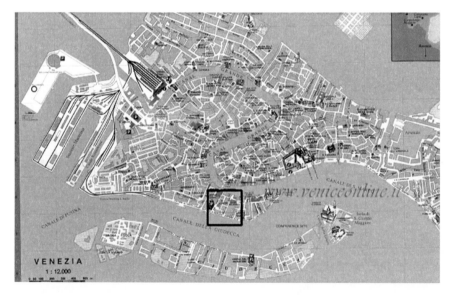

Fig. 6.24 Location of the studied area in the southern part of Venice in the Sestiere of Dorsoduro

geographical surface of 285 m × 305 m = 86,925 m². The total surface of the 279 façades measured is 48,361 m²(Table 6.6).·

The types of buildings present in this area are:

- *Palaces,* mainly on the Canal Grande: Palazzo Brandolin, Palazzo Contarini dal Zaffo (15th Century), Palazzo Balbi-Valier, Palazzo Loredan, Palazzo Barbarigo (16th Century), Palazzo Da Mula-Morosini (15th Century), Palazzo Centani, Palazzo Venier dei Leoni (uncompleted: only the first floor was constructed) (Fig. 6.25). In the centre of the studied area there are some small palaces like the Palazzetto of the 15th Century situated n° 832–834 Piscina Venier.
- *Churches*: Spirito Santo (1483) and Sant'Agnese (13th Century) (Fig. 6.26).
- *Museums:* Gallerie dell'Accademia (Campo Santa Maria della Carita), Galleria Cini (Campo San Vio) and Guggenheim Collection (in Palazzo Venier dei Leoni).
- *Great architectural complexe*: ex-Ospedale degli Incurabili (16th Century).
- *Typical traditional private houses*, especially in the quarter delimited by the Rio delle Toresele (W), the Fondamenta Bragadin (N), the Rio della Fornace (outside the Eastern limit of the studied area) and Fondamenta Zattere allo Spirito Santo (S) (Fig. 6.27).
- *Modern buildings* are rare: Casa Cicogna (1954–1957) (Fondamenta Zattere allo Spirito Santo, Dorsoduro 401).

Two small canals run across the studied area: the Rio di San Vio and the Rio delle Toresette. Their bridges (stone, mortar and brick), as well as the decorated wells (stone) of the "campi" and the stone of the quays, were not taken into account due to difficulties in evaluating their developed surface in the field.

Fig. 6.25 View of the southern bank of the Canal Grande, Venice, taken from the Accademia Bridge with the Santa Maria della Salute Church in the background. The palaces on the right determine the northern limit of the studied area

Fig. 6.26 Façade of the Sant'Agnese Church, Campo Sant'Agnese (S = 158 m²: 5% stone, 95% brick)

The major materials constituting the façades are: stone (the white Pietra d'Istria), rendering-mortar and brick. The renderings are applied onto brick (except for the modern Casa Cicogna where rendering is applied on concrete) which reappears when render detaches. Metals (grids, portals) and wood (doors, windows, and pillars) are present but their surface was not estimated. Important mosaics in coloured glass decorate the façade of the Palazzo Barbarigo on the Canal Grande.

The risks for these materials in Venice are:

- *water rising* (containing dissolved sea salts) at the base of walls, causing damage on several meters height. The crystallization of salts when water evaporates leads to the detachment of renderings and mortars, the desegregation of stone and brick.
- *"acqua alta" (high waters):* periodic floods due to the conjunction of several phenomena: high tide, southern wind, atmospheric depression and the shape of the laguna. The inundation of places and streets causes humidity and major sea water rising with deposition of sea salts on the materials.

Fig. 6.27 Typical traditional Venetian popular house, Ramo primo agli Incurabili, Dorsoduro n° 436. (S = 96.80 m²: 100% render). Originally the entire brick surface was covered by the render which detaches at the basis of the walls due to capillary raisings and salt crystallisation (white efflorescences)

- *air pollution* from industry (in the complex of Marghera-Mestre), traffic (vaporetti, taxi-boats, boats for transportation of merchandises) and domestic heating (methane) often results in the formation of gypseous black crusts and soiling of surfaces.
- *surface recession and mass loss* by erosion-dissolution in the parts of stone and mortar/rendering exposed to the rain.
- *air humidity* associated with marine aerosols (sea spray), due to the situation inside the laguna of Venice, causing permanent wetting of the surfaces and favouring biological colonisation (lichens, mosses, algae).
- *mechanical action* of waves produced by the very high speed of motor boats.
- *mass tourism* (graffiti, litter, pigeon excrement): intense in some restricted areas in Venice (e.g. Saint Mark's Place, Rialto Bridge) but located outside the studied zone, although the presence of the Guggenheim Collection in Dorsoduro attracts many people but perhaps more cultural and more educated than mass tourism.

The methodology employed for the setting up of the inventory of materials in the façades of Venice-Dorsoduro is based on the direct measurement of the length (L) and height (H) of each building or monument using a laser beam (Professional Digital Laser Distance Measuring Instrument DLE 150 Bosch). The surface of the façade was then calculated (S = L × H). The proportion (%) of each type of material (stone, mortar/rendering, brick) was visually estimated and its surface easily deduced. An example of laser measurement results (Length and Height) and calculation sheet of the total surface of façades and percentage and surface of the different constituting materials performed in the Calle delle Mende is given in Table 6.6.

The evaluation of the stock of materials at risk present in the façades of Venice-Dorsoduro displayed below results from 558 measurements (L and H) performed on 279 buildings. Its statistical representativity is thus very high.

The total surface of façades (48,361 m^2) consists in 4,808 m^2 of stone, 36,846 m^2 of rendering and 6,676 m^2 of brick (Table 6.6).

The stone (the white Pietra d'Istria) occupies a mean value of 10% of the total surface of the façades in Dorsoduro. This material is thus in the minority, except on the Canal Grande where it represents 43% due to the most decorative architecture of the Palaces.

Rendering and mortar represent a mean value of 76% of the total surface of the façades (51% on the Canal Grande); represent by far the most used materials in Venice façades. It has to be kept in mind that rendering and mortar are generally applied on walls of brick, with the latter reappearing when render and mortar detach; this is the case at the base of all the walls due to capillary risings and salt crystallisation.

The mean value for brick is 14% of the total surface of the façades in Dorsoduro when remaining naked (not originally covered by render or mortar). There are no brick façades on the Canal Grande, where there are 6% of glassy mosaics.

6.3.3 The Stock of Materials at Risk on the Façades in the Via del Babuino in Rome

The centre of Rome, inside the Urban VIII walls, is inscribed on the UNESCO List since 1980 because of its exceptional universal value, which is universally recognized.

For the evaluation of the Stock of Materials at Risk in Rome, a typical and representative area was chosen: the Via del Babuino joining the Piazza di Spagna to the Piazza del Popolo (Fig. 6.28). It is representative of the civil buildings of the historical centre of Rome and comprises three churches: Sant' Anastasio, All Saints and Santa Maria di Monte Santo (at the corner of Piazza del Popolo) (Fig. 6.28).

The employed methodology for the setting up the inventory of materials in the façades of Via del Babuino is the same as in Venice – Dorsoduro (see above).

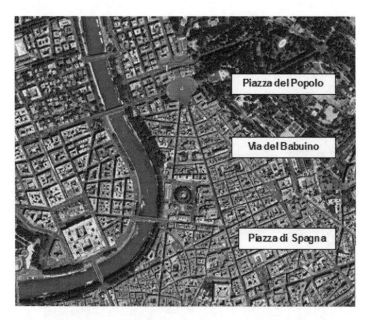

Fig. 6.28 Map of the centre of Rome showing the Via del Babuino between the Piazza di Spagna and the Piazza del Popolo

The 70 measured façades spread for a total length of 921 m alongside the street (Table 6.6).

Besides the three churches, the buildings of the Via del Babuino are ancient medium sized palaces converted into private flats with high standard shops at the ground level (fashion, design, antiques, and restaurants). The façades observation reveals that they were painted on a very large scale during the restoration and cleaning works performed for the Jubilee and the new Millennium in 2000. All types of materials were painted and probably not cleaned before this treatment, inducing an uncertainty for their future behaviour. Thus, many columns in travertine are half painted at the ground level with the aim of conservation of the aesthetics of the original stone (Fig. 6.29).

The total surface of façades (16,913 m^2) consists in 3,411 m^2 of naked travertine (20%), 338 m^2 of volcanic tuff (2%), 42 m^2 of marble (0.2%), 902 m^2 of naked rendering (5.4%) and 739 m^2 of brick (4.4%). The painted materials are predominant comprising 11,479 m^2, or 68%, of the total surface of the façades (Table 6.6).

In contrast, two churches (Sant'Anastasio and All Saints) contain 85% and 90% respectively of naked brick, and only 15% and 10% respectively of travertine. The third church (Santa Maria di Monte Santo, Fig. 6.30) presents on the Piazza del Popolo a frontage containing 90% of travertine and 10% of painting, and on its side along the Via del Babuino only 5% of travertine but 95% of painting. In Rome, the most important monuments are in marble and travertine, while the façades of other buildings were in coloured rendering before their recent painting in 2000.

Fig. 6.29 Two examples of columns half painted for the jubilee and new millennium in 2000: Via de Babuino n° 110–118 and n° 121, Rome

Fig. 6.30 Frontage on the Piazza del Popolo (90% travertine, 10% painting) and side on the Via del Babuino n° 199 (5% travertine, 95% painting) of the Santa Maria del Monte Santo Church, Rome

The risks for these materials in Rome are double: traffic (soiling) and mass tourism (degradation). Another risk was induced by the recent painting applied on all materials and particularly on dirty deposits and black crusts without any previous cleaning or soft sandblasting. The rapid detachment of the paint is inevitable in the near future.

6.3.4 Conclusions from Case Study

The Direct Measurement Method applied to the façades in the centre of Paris, the Sestiere of Dorsoduro in Venice and the Via del Babuino in Rome allowed to point out the very strong contrast between Paris, Venice and Rome regarding the constituting materials (Table 6.6): 76% in Paris versus 10% in Venice and 20% in Rome for limestone/travertine/marble, 7% versus 76% and 5.4% respectively for rendering/mortar, 15% versus 0% and 68% respectively for painting, and 1% versus 14% and 4.4% respectively for brick. In the three cities, metals play a minor role in the façades. When regarding the most important historical monuments, in Paris they are entirely (100%) in Lutetian limestone while in Venice-Dorsoduro the palaces located on the Canal Grande contain 43% of Pietra d'Istria, 51% of rendering and no brick in their façades, but relatively important glassy mosaics (6%). In Rome, two churches located in Via del Babuino are made of 10–15% of travertine and 85–90% of brick without any painting, while the third church at the corner of the Piazza del Popolo shows a strong contrast between its side and its frontage.

Paris is a city of stone, Venice a city of rendering applied onto brick and Rome is a city of painted materials, essentially travertine and rendering.

The proportions of the different materials used in the façades of buildings and monuments in the core of historical cities are intrinsic characteristics of each city: they are determined by the presence of sites of extraction (quarries) in the immediate environment of the city since stone is a heavy material and its transportation is very expensive. The different materials used in different cities also depend on local history and cultural, architectural and handicraft traditions, but also on methods and materials used for maintenance and restoration works.

Acknowledgments The authors of this chapter wish to express their appreciation to all the colleagues in the MULTI-ASSESS and CULT-STRAT consortia for their support, provision of data and stimulating discussions. We are particularly grateful to Fernando Viejo, Jesús M. Vega and M. Morcillo from the National Centre for Metallurgical Research (CENIM/CSIC), Madrid, Spain, Edward Andrews, Middlesex University, UK, Jan Bryscejn, ITAM, and Dagmar Knotková from SVOUM, Czech Republic for their contributions to this chapter.

References

Andrews, E. (2006). Auditing of stock-at-risk of materials of the built heritage: a case study of the city of London churches. MSc Dissertation, MSc Risk Management, Middlesex University.
ApSimon, H.M. and Cowell, D. (1996). The benefits of reduced damage to buildings from abatement of sulphur dioxide emissions. *Energy Policy* 24(7):651–654.

Cobb, G. (1962). London City Churches- A Brief Guide. The Corporation of London

Fedeski, M. and Johns, J. (2001). Assessing the impact of climate change on the building stock of a region: Identifying an appropriate methodology. In: International Scientific Meeting on the Detection and Modelling of Recent Climate Change and its Effects on a Regional Scale, May 2000 Tarragona, Spain, pp. 617–633.

Haagenrud, S. (1997). Assessment of working life of buildings and building products through characterization of degradation environment and modelling of degradation. In: 1st Regional Modelling of Air Pollution in Europe Workshop, Date: SEP 26–27, 1996, Copenhagen, Denmark, p. 265.

Hatts, L. and Middleton, P. (2003). London City Churches. Bankside Press. London.

Yates, T., (1998). Use of Identikits for Assessment of Stock at Risk. In National Materials Exposure Programme Part 3: Economic Evaluation of the Effect of Air Pollution on Buildings and Building Materials. Prepared for Peter Woodhead CIRM, DETR. Building Research Establishment Ltd 1998.

Sources of Additional Information

Some details of the Risk Map of Cultural Heritage in Italy can be found here:
http://www.uni.net/aec/riskmap/english.htm

The defining of the character of landscapes discussed above is being undertaken by English Nature and The Countryside Agency and further details can be found on the Countryside Agency's website

(http://www.countryside.gov.uk/LAR/Landscape/CC/index.asp)

More details of the objects described in the case study can be found at:

Paris

http://whc.unesco.org/en/list/600.

Venice

http://whc.unesco.org/en/list/394).

Rome

http://whc.unesco.org/en/list/91).

Chapter 7
Economic Evaluation

John Watt, Ståle Navrud, Zuzana Slížková, and Tim Yates

7.1 Overview

The preceding chapters have, to a large extent, concentrated on an examination of the scientific methods of working out how much heritage materials are damaged by the environment that they are in. Science alone, however, does not tell us what to do about it. In the real world, the decisive factor is usually cost. There are many influences on the cost that we are prepared to accept for the given goods or services and many of those are linked to the value that we place on things. This chapter looks at some of the economic dimensions of air pollution damage to heritage.

The first research reviewed is the relatively straightforward, at least in principle, task of calculating real costs for maintenance and repair, which are an important category of information needed for the calculation of direct costs using the dose-response/damage function approach. Several pieces of information are required to make an estimation of the costs associated with this damage to cultural heritage.

- Measurement of environmental parameters, including pollution
- Selection of dose/response and damage functions
- Estimation of the amount of each material used
- Establishment of critical levels of damage
- Estimation of the costs of repair or replacement

The first three topics have already been discussed and the remaining two will be presented in later chapters so it is possible, for now, to concentrate on the estimates of the costs of repair or replacement. A number of studies that have tried to collate real costs have shown that prices can differ depending on the size of the object, the extent of damage and of the work and the choice of contractor but may include some or all of the following categories.

J. Watt (✉)
Centre for Decision Analysis and Risk Management, School of Health and Social Sciences, Middlesex University, The Burroughs, London NW4 4BT, UK
e-mail: j.watt@mdx.ac.uk

J. Watt et al. (eds.), *The Effects of Air Pollution on Cultural Heritage*, 189
DOI 10.1007/978-0-387-84893-8_7, © Springer Science+Business Media, LLC 2009

- material cost
- labour including management costs
- architect/engineering work
- removal from building
- transport and storage of waste

For objects of cultural heritage higher prices can in general be expected as the demands are often higher with respect to professional skill, quality of material, complication of design, artistic elements etc. Nonetheless it is possible to make an estimate of what a given amount of damage will cost to repair.

The benefit of this may not be immediately obvious to somebody with a heritage background. After all, wouldn't it be simpler to survey the damage visible on the building and get a series of quotations for the repair? The value lies in the ability to estimate the fraction that results from air pollution, which has two potential benefits. First, and most importantly perhaps, it can be combined with stock-at-risk estimates and damage maps to develop regional cost estimates for heritage materials to inform air quality policy development. Cost maps can also be developed which highlight priority areas for action. Second, knowledge of the likely progress of air pollution damage and its associated costs allows the heritage protection sector to plan for future maintenance.

Another important part of the economics of heritage conservation is an estimate of the value that people place on repair and cleaning. This gives the other side of the cost/benefit equation. One part reveals costs and the other part calculates the level of resource available for conservation and repair. Again, local information may be acquired in a different way – a repair is needed and an appeal launched – but study of the underlying economics can be very revealing.

Cost-benefit analysis is a method that has been developed to evaluate the desirability of investing in a project by establishing its (net present) value after discounting its costs and benefits to society. If the value is positive, investment may be considered worthwhile. The main difference to ordinary investment appraisal used in commercial accountancy by companies deciding whether to invest in something or not (where only monetary values and outlays are considered), is that cost-benefit analysis seeks to evaluate all phenomena that can be said to have a utility or disutility. (The concept of utility is an economic term crucial to the analysis of the behaviour of individuals, usually defined as the satisfaction that individuals gain from buying products (whether goods or services). The assumption is that individuals will make choices that maximise their utility.). Thus a cost-benefit analyst appraising a cathedral building project would include as costs not only the financial outlays involved in the construction, maintenance and repair of the project but also disutilities such as reduced access and reduction in visitor numbers and noise nuisance. The benefits side would not only include increased revenue subsequent to the project but also the increased security of the object for future generations and the increase in aesthetic appeal. A commercial company would have no interest in such social costs and benefits (except perhaps the extent to which they affect employee performance).

Since social costs and benefits are measured in many different types of unit (time, decibels, micrograms of pollution etc.), it is necessary to reduce them to a common unit of account – namely money. When the concept of a market is applied to issues involving cultural heritage goods, two problems immediately emerge. The first is in defining what is meant by "the goods" that are being "traded" in the market. Here, the goods being valued are interventions or actions that are undertaken to maintain our heritage. The concern is not with valuing the intrinsic worth of the object itself.

The second problem is that many such goods are not traded in markets and therefore don't have any observable price. Cultural heritage goods are typically thought of as public goods, which can be valued by a number of techniques that also contain the concept that people may value heritage even though they don't visit it. This leads to what are known as non-use values, which are often very important in undertaking a cost-benefit analysis, since they represent potential income (or at least are an estimate of benefit).

There are a number of non-market valuation techniques that can be used to estimate a monetary value for cultural heritage goods, drawn in large part from environmental economics. Only a relatively few studies have directly examined valuation of heritage and this chapter presents an overview, to point out the unique challenges and opportunities involved. The chapter also briefly examines the potential impact on the local economy.

7.2 A Consideration of the Use of Cost-Benefit Analysis in Relation to Cultural Heritage

Cost-benefit analysis has been perceived by some people to be a somewhat controversial technique when applied to cultural heritage. It is important to stress that the focus is on rationalising pollution damage by setting it in the widest possible context of valuation, cost estimation and examination of policy. This, therefore, is not just a cost benefit analysis. While accepting that economics provides a useful structure to the balancing of different effects, it should be stressed that there are pitfalls, usually relating to underlying assumptions or inappropriate establishment of economic values to social costs. It seems appropriate to discuss the cost benefit technique and its likely applications within the heritage sector.

Cost-benefit analysis is beginning to be widely applied in governmental planning and budgeting and describes the attempt to measure the social benefits of a proposed project in monetary terms and compare them with its costs. A cost-benefit ratio is determined by dividing the projected costs of a programme by the projected benefits. This is valuable in setting priorities for investment since a programme with a low ratio can take priority over others with higher ratios. Determining this ratio is usually a difficult task, however, because of the wide range of variables involved. Both quantitative and qualitative factors must be taken into account, especially when dealing with social programmes. For instance, the monetary value of the presumed benefits of a given programme may be indirect, intangible, or projected far into the future. The time factor must be considered in estimating costs, especially in long-range planning.

In the United States cost-benefit analyses have been used since the 1960s, in all aspects of government planning and budgeting, from programmes that can be analysed with reasonable accuracy, such as waterworks, to programmes that involve a great degree of subjective data, such as military strategy.

Critics of cost-benefit analysis argue that reducing all benefits to monetary terms is impossible, and that a quantitative, economic standard is inappropriate to political decision making. Calculation of what are known as "indirect costs" (see below), may give the greatest cause for concern, since the existing studies are few in number and the ability to transfer findings between different situations is very controversial. Concern has been expressed that different cultural objects would be forced into unreasonable competition with each other with invidious comparisons being made as to their intrinsic value. What is not in doubt is that society values heritage for historic, cultural and aesthetic reasons but, often, attempts to place a monetary value on these benefits are not seen as desirable.

The counter argument is that in every society funds are limited to some extent and only economics gives a sufficiently rigorous structure within which to attempt an analysis of the relative merit of different proposals. Exponents of the technique are at pains to emphasise that the economic valuation should focus on techniques to value the effect of *interventions* on the state of heritage objects not on valuing their intrinsic worth as objects. Owners or managers of listed buildings that would be worried about any attempt to rank them as objects, ought to be reassured by a technique that allows them to assess, for example, whether a cleaning programme will be of overall benefit to their fabric or not.

In addition, there is no doubt that implicit cost-benefit analyses are undertaken constantly anyway, including those which decide the outcome of funding applications. Heritage projects are in direct competition with each other in many cases and it would be useful to have a tool which facilitates the optimum presentation of all of the supporting evidence. Careful valuation studies will also reveal the best strategies for raising funds, which may highlight important additional sources of finance that are currently being under utilised.

When national and international governments need to examine the value of cost-benefit analysis focussing on damages arising from air pollution, it is clear that this analysis needs to include heritage effects as well as human health impacts. The benefits to be gained from reductions in pollution need to be properly assessed if they are to justify the, much easier to quantify, costs of remediation. Much of the damage is irreversible and robs future generations of a valuable legacy. In principle, estimation of monetary values for external effects allows for adjusting the market signals such that these reflect such social costs. Monetary valuation of the external effects can be included within a cost-benefit analysis-based approach to appraisal plans for cultural heritage protection, which would provide an improved decision-making basis taking into account the widest possible set of impacts.

In some cases it is argued that the cultural heritage is invaluable, or has an "infinite value". In that case the priority of action would simply be based on a comparison of cost of pollution reduction. However, often semi-quantitative rankings can be used to make lists of priority for action. It may also be possible to put a monetary value on the degradation, e.g. as the cost of the corresponding maintenance. General recommendations to implement air pollution reduction policies to protect cultural heritage are based on descriptions of typical stock at risk and typical emission sources, concentrations and exposure effects of the pollution, with particular focus on hot spots. Policy advice would go on to describe the typical estimated cost of the pollution in terms of degradation effects and related maintenance needs, and then give general recommendations for pollution reductions. A description on this level, with quantified examples together with measurements of local conditions, can be a good basis for recommendations of action to be taken. The starting point could be monitored concentration values for pollution, as described earlier and for which data is available in many cities. Recommendations would then focus specifically on the reductions in concentration levels required to achieve the policy objectives, with suggestions for the measures most likely to achieve the lower concentrations, based on general knowledge about the typical sources and dispersion processes.

A further step towards targeted recommendations would be to present reduction scenarios and utility calculations for selected case studies at regional and local scale, e.g. in selected cities based on emissions – dispersion – abatement modelling. If such a modelling system is available for the region or city it is possible to give better-founded and more specific recommendations for action. Whether recommendation are based on measurements or emission-dispersion modelling it should preferably also include emissions (or concentration) – effects and cost-utility calculations.

7.3 Calculation of Direct Cost for Maintenance and Repair

An important category of information needed for the calculation of direct costs using the dose-response/damage function approach is the real costs for maintenance and repair (a number of activities can be important depending on context – resulting in different categories of cost for work on the fabric of a building – such as conservation costs, renovation costs, repair costs and maintenance costs). The prices can of course differ depending on the size of the object, the extent of damage and of the work and the choice of contractor. The prices may include some or all of the following categories:

- material cost
- labour including management cost
- architect/engineering work
- removal from building
- transport and storage of waste

For objects of cultural heritage higher prices can in general be expected as the demands are often higher with respect to professional skill, quality of material, complication of design, artistic elements etc.

Average European conservation and renovation costs are calculated from data supplied from Czech Republic, Italy, Norway and United Kingdom and presented in Table 7.1 (CULT-STRAT, 2007). This sample is not sufficiently large for any statistical evaluation and serves rather as a representation for very basic considerations of the pollution cost range and diversity conditions related to cultural heritage, (see below). In addition to the data presented for roof envelope costs and typical lifetimes, data have also been collated for non-plastered masonry, glass walls, plastered façades, painted wood, supporting works and sculptures or sculptural items.

Table 7.1 Average European conservation and renovation costs

No.	Surface layer or system	Unit	Guide cost EURO			Lifetime in years		
			Minimum	Maximum	Average	Min	Max	Average
1	Copper sheet – total replacement	m^2	55.65	85.00	67.82			
2	Steel galvanized sheet – total replacement	m^2	16.04	52.00	30.59	5	45	25
3	Plain tiles (double) – total replacement	m^2	30.23	43.32	36.58	40	100	70
4	Flap pantile roofing – total replacement	m^2	24.23	45.39	37.65	30	70	50
5	Wooden shingle split – total replacement	m^2	35.16	83.68	58.04	50	80	65
6	Paint on steel sheet – renewal	m^2	3.75	5.48	4.62	3	9	6
7	Paint on galvanized steel sheet – renewal	m^2	5.53	57.00	26.02	5	11	8
8	Paint on wooden shingle roofing	m^2	5.32	43.32	17.96	5	8	6,5
9	Regional variants – slate tiles	m^2	51.88	53.61	52.74			
10	Flap pantile roofing – repair and making up to 10% of tiles	m^2	3.99	133.68	49.30			

Data supplied from Czech Republic, Italy, Norway and United Kingdom, CULT-STRAT, 2007

7.3.1 *Assumptions*

Substantial maintenance is carried out on exterior surfaces of buildings since the materials making up the frontage undergo a continuous process of degradation. This process can, for practical purposes, be split into two main components: natural degradation which takes place irrespective of air quality, and degradation due to emission of pollutants into the atmosphere. Figure 7.1, based on the dose-response functions introduced in Chapter 3, shows there is a measurable level of mass loss even where SO_2 values are zero. When calculating costs of corrosion of building materials it is not enough to consider the actual cost of maintenance work. The maintenance that would have been required in zero-pollution environment ought to be subtracted to obtain a correct estimate of material costs due to air pollution.

Another assumption is that the owner of the building is him/herself in a position to decide when maintenance is to be carried out. A reasonable assumption is that owners wish to minimise total outlays on maintenance of the building over time and will therefore actively adapt maintenance frequency to maintenance needs. Calculations in the REACH programme (REACH Project: "Rationalised Economic Appraisal of Cultural Heritage" ENV4-CT98-0708), which examined the economics of air pollution damage to heritage, incorporated two main assumptions about agents' information and patterns of behaviour. First, it was assumed that the owner of the building knows when maintenance of the building should be performed in order to minimise costs in the long term and that he/she performs the work at this

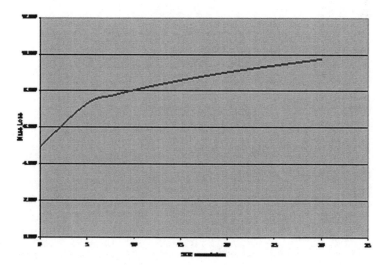

Fig. 7.1 First year exposure, Cu mass loss as a function of SO_2 concentration calculated from a dose-response function

point. Second, aesthetic motives behind a good deal of maintenance work were disregarded. If the agent performs maintenance more often than required by the criterion of physical wear and tear, which is often the case with private individuals, air pollution may not alter agents' maintenance frequency. If this frequency is higher than optimal maintenance frequency based on technical criteria, air pollution will not result in increased costs for the owner. On the other hand, the long-term costs may exceed such calculations for owners with basic maintenance frequencies who do not adapt this frequency to the increased wear and tear caused by air pollution. Information from the construction industry suggests that agents are more concerned with adapting building maintenance to budget constraints than to maintenance needs. The sum total of these sources of uncertainty resulting from lack of information gave no clear indication as to its effect on costs.

7.3.2 Model for Calculating Costs due to Air Pollution

The REACH project defined the following variables for material corrosion costs:

K_b = Material corrosion costs in the background atmosphere per year
K_f = Material corrosion costs due to national/local air pollution per year
K_t = Total material corrosion costs per year

They obtained:

$$K_t = K_b + K_f \qquad (7.1)$$

REACH further defined:

M = stock of a given type of material at risk (m^2)
t_b = lifetime of this material in the background
      atmosphere (years)
t_t = actual lifetime for the material (years)
P = maintenance cost per square metre of the material
      (in local currency)

Thus for each type of material:

$$K_b = \frac{M \times P}{t_b} \qquad (7.2)$$

$$K_t = \frac{M \times P}{t_t} \qquad (7.3)$$

Equation 7.1 gives:

$$K_f = K_t - K_b \tag{7.4}$$

Inserting Equations 7.2 and 7.3 in Equation 7.4 gives:

$$K_f = M \times P\left(\frac{1}{t_t} - \frac{1}{t_b}\right) \tag{7.5}$$

The variables in this equation will have different values according to the material studied and the nature of the air pollution in the area in question. Total corrosion costs due to pollution for any given area would therefore represent the sum of the calculations for each material and each pollutant.

7.3.3 Estimation of Air Pollutant Dose–Exposure Cost for Selected Stock at Risk of Cultural Heritage

The total utility of reduced degradation of the cultural heritage, when air pollution is reduced, can be estimated as:

Utility of reduced air pollution degradation of the cultural heritage = utility of action to reduce the pollution (in terms of reduced degradation) – cost of pollution reduction.

The total utility is calculated as the quantifiable savings in terms of reduced material degradation for the stock at risk and minus an estimate of the pollution reduction cost. The actual loss of value of cultural heritage due to its degradation in the environment may be difficult to quantify. One approach is to simply calculate the cost of the best possible physical maintenance of the heritage as set out here.

A simple formulation for the calculation can be expressed as:

A estimation of the total area of stock at risk (m^2)
B estimation of the cost of repair, expressed as (cost in $/m^2$ surface treated)
C estimation of the acceptable level of corrosion a as discussed in Chapter 9, ($\mu m\ yr^{-1}$)
D the value from the Dose-Response function ($\mu m\ yr^{-1}/\mu g\ m^{-3}$)
E the existing air pollution concentration ($\mu g\ m^{-3}$)

Then cost of degradation $= \frac{A \times B \times D \times E}{C}$

7.4 Methods for Valuing Welfare Loss of Damages to Cultural Heritage (Indirect Costs)

The *welfare loss* of damage to cultural heritage by pollution (or indeed any other damage) is a term that refers to the additional losses that are added to the repair and maintenance costs discussed in the previous section. These additional costs usually result from a reduction in the number of visitors (and thus a reduction in income) together with some measurable loss of the value of their experience. This might result from a cherished building being covered in scaffolding for example or a peaceful cathedral echoing to the sound of pneumatic drills, both of which might lead to a visitor getting less out of their visit or postponing it altogether. Such losses, though real, are much harder to quantify than the direct repair costs since they are not traded and so do not have an obvious price. Nonetheless, they are important since, if a reasonable estimate can be obtained of the amount of money that people would pay to avoid the loss (or be willing to pay to rectify the situation), this value can be used as an estimate of the budget potentially available to deal with the problem. Obviously willingness-to-pay surveys do not provide an actual income but can be used as evidence of the benefit of a repair, maintenance or conservation activity as part of a funding bid.

A number of approaches towards the valuation of non-marketed goods have been developed in recent years including:

Implied market decisions revealed preferences (RP);

- hedonic pricing;
- travel cost method (TCM);

Experimental market techniques

- stated preferences (SP);
- contingent valuation method (CVM);

Surrogate market methods

- replacement cost method;
- shadow prices;
- surrogate markets.

These methods have been reviewed with particular reference to cultural heritage (Navrud and Ready, 2002). The most useful technique is the contingent valuation method (CVM), which is a questionnaire-based valuation technique whereby willingness to pay (WTP) or willingness to avoid (WTA) are directly obtained from the respondents with respect to the specific goods (or action). CVM has mainly been used with respect to non-marketed goods and in particular for environmental goods/resources. In this context a hypothetical market for non-marketed goods is defined and the respondent is requested to specify their WTPs (or WTAs).

In persuading the sceptical about the potential of cost-benefit analysis, it is useful to remind them that a principle advantage lies in permitting the user to capitalise on the area of benefits. The user is able to define the extent of potential benefits based on their willingness-to-pay, and then he can seek to "capture" these benefits.

To begin with, the beneficiaries must be defined.

As an example, for Cathedrals benefits may be derived from:

- direct use of the resources (such as visitors and religious users)
- indirect use (seeing pictures of the cathedral buildings or listening to music recordings from its choir)
- Non-users (these may not have anything to do with specific uses but may relate to pleasure that the object exists (existence value) and will be available for others (altruistic value) and future generations (bequest value) or for possible future use (option value). These values may be a significant proportion of the total value of cultural heritage. They also may be relatively hard to capture in terms of direct monetary input, extending as they do outside national boundaries and current generations.

These different types of beneficiaries have different priorities, different impacts and different potential for accruing benefit. These important characteristics are established by survey.

Cost-benefit analysis can also be useful in making the choice between a number of strategies. Figure 7.2 illustrates a situation where the entrance fee controls the number of visitors. The dotted lines show two possible choices that a cathedral might make – many visitors with a low yield per visitor or few

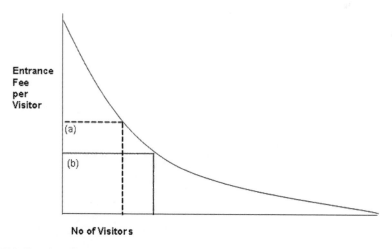

Fig. 7.2 Cost-benefit analysis for making the choice between a number of strategies in a situation where the entrance fee controls the number of visitors (Source: Navrud in the REACH Project).

visitors, each paying a substantial amount. Either extreme is likely to be less than ideal, and any strategy would look to defining an acceptable trade-off. Economic analysis would be required to establish the relationship between price and attendance, as well as to work out the optimum level of income. In the illustrated example, the two strategies give the same income but the price had an effect on attendance. A cathedral that is often crowded might adopt one measure (a) to maintain tranquillity and a cathedral that derives a lot of income from souvenir sales might adopt the other (b).

The welfare loss of damages to cultural heritage from corrosion can be divided in two main categories:

- welfare loss to visitors and non-visitors of the cultural heritage site
- lost local income and employment impacts (including spin-off or multiplier impacts) from reduced number of visitors to the site

In a social cost-benefit analysis of the restoration of a historic building, for example, only the first category is relevant, as the second category in most cases would create zero social benefits, since it cannot be ruled out that the local income and employment impact of visitors to the cultural heritage site could have been created with a completely different local project costing less than the restoration project (and/or the same restoration project done at another site would have created larger local income and employment impacts). Therefore, if the level of decision-making is the national level, only the welfare loss to all domestic visitors and non-visitors of the cultural heritage site should be counted. If the level of decision-making is local, e.g. a community, only welfare losses to households in the community should be counted, but then all direct losses in local income and employment and multiplier effects should be added, see below.

The following simple model has be used to assess the magnitude of the social welfare loss from changed appearance of a cultural heritage object due to corrosion, i.e. category (i) above:

$$\Delta TEV = \Delta V + \Delta NV$$

The welfare loss will be the change in the annual Total Economic Value (ΔTEV) due to the changed condition/appearance of the cultural heritage object from corrosion. ΔTEV is the sum of change in annual use and non-use value of visitors (ΔV) and the change in annual non-use value of non-visitors (ΔNV) due to the changed appearance of the cultural heritage object.

The change in annual use and non-use value among visitors is equal to the mean willingness-to-pay (WTP) per visitor (or visitor day) to experience the site times the change in annual visitation rate (measured as annual number of visitors or visitor days), ΔnV, due to the change in the appearance of the cultural heritage object. We have assumed a constant WTP per visitor, independent of the condition/appearance of the cultural heritage object. This assumption can be relaxed when we have studies that show how WTP for visitors vary with the appearance of the object.

$$\Delta V = \Delta WTPV \times \Delta nV$$

The change in value to non-visitors is a product of the mean annual WTP per household to restore or preserve the cultural heritage object from the corrosion damages (WTPNV), and the number of households experiencing a welfare loss (Δ nNV) from corrosion done to this object.

$$\Delta NV = \Delta WTPNV \times \Delta nNV$$

7.4.1 Methods for Valuing Cultural Heritage Interventions

Before it is possible to measure the value of the goods, any goods, the concept of "value" must be defined. Here, a very neo-classical economic perspective can be adopted, where the value of the goods is defined as either:

- the amount of money that the potential consumer would be willing to pay to get the goods (willingness-to-pay or WTP), or
- the amount of money that the owner of the goods would have to be paid in order to induce him or her to part with it (willingness-to-accept or WTA).[1]

Where goods are actually traded (sold), consumers can compare their own value for the goods (their WTP) with the market price, if their value for the goods is larger than its price, they will purchase it. Owners (or producers) of the goods can compare their value for the goods (their WTA) with the market price, if their value for the goods is less than the market price, they would sell. Economists can therefore say that, "in market equilibrium, the marginal consumer and the marginal producer will have WTP and WTA respectively for the goods that is just equal to its price. For that reason, the market price can be seen as a signal of the good's marginal value to both consumers and providers".[2]

Application of this conceptual framework to issues involving cultural heritage goods, leads to two immediate problems. The first is in definition of "the goods". As discussed above, it is rarely useful to consider a heritage object itself as the goods to be valued. Rather, it is often more useful to value some change in the characteristics of the building or monument, for example a change in its

---

[1] It should be noted that this approach gives complete sovereignty to the individual. Goods have value only because individuals want them, and are willing to trade money in order to get them. In particular, this definition of value rules out intrinsic values for goods that are independent of people's preferences for those goods.

[2] Navrud and Ready (2002). Note also that price may also represent the goods' marginal value to society depending on whether there are any externalities in consumption or provision, that is any impacts from its consumption on individuals other than the seller and the buyer.

appearance. If Q^0 represents a physical description of the building or monument (e.g. "dirty"), and Q^1 represents some different physical description (e.g. "clean"), then the goods to be valued can be defined as the difference between Q^0 and Q^1. For example, the soiling of a cathedral from airborne pollutants is being evaluated; Q^0 could represent a more-soiled facade, while Q^1 represents a less-soiled facade.[3]

The individual whose value is being estimated has some utility function that ranks different sets of outcomes. This utility function will logically include as an argument the individual's wealth or income, I, as well as the physical state of the cultural heritage goods, Q, so that the individual's utility is given by V(I, Q). Combinations of I and Q that yield higher levels of utility are preferred to combinations that yield lower levels. The value that the individual places on the change from Q^0 to Q^1 (in this case the individual's WTP for a change from Q^0 to Q^1) is then the largest amount of money that the individual would willingly give up in order to achieve Q^1 instead of Q^0, and is given by[4].

$$V(I, Q^0) = V(I - WTP, Q^1).$$

It is this measure of value that the economist seeks to estimate empirically.

The second difficulty is of trying to apply the market model to cultural heritage goods, namely that they are rarely traded in markets, and thus have no observable price. There are very good reasons why cultural heritage goods are not traded in private markets. Cultural heritage goods are typically public goods, meaning they have two precisely-defined characteristics. First, the benefits (values) generated by cultural heritage goods are typically non-rival, that is the benefit enjoyed by one individual does not come at the expense of the next individual's enjoyment. This is in contrast to market goods, where a given unit of the good can only be consumed by one individual. Second, it is often difficult to force people to pay a price before they can enjoy the benefits from the cultural heritage goods. Even where entrance to a building can be regulated by an entrance fee, the non-user benefits accrue regardless of whether they have been paid for. In formal economic terminology this means that the goods or that enjoyment of the goods is non-excludable.

These two conditions lead to a situation where markets cannot be trusted to provide an adequate supply of cultural heritage goods. It is for this reason that such goods are usually provided collectively, either by governments or by groups of people working co-operatively.

The absence of a price means that values for cultural heritage goods cannot be observed directly. Instead, clues must be sought that reveal something

---

[3] In extreme cases, Q^0 could represent the complete loss of the building or monument, but complete elimination of a building or monument is rarely policy-relevant.

[4] In the economics literature, this measure of value is called the individual's compensating variation for the change, or less precisely their consumer surplus.

about value indirectly. Non-market valuation is a term used to describe a variety of techniques for looking for and interpreting these clues about value for goods that are not traded in markets. There are two broad categories of non-market techniques: revealed preference techniques and stated preference techniques. As the name implies, revealed preference techniques involve searching for those clues by examining an individual's past behaviour. One type of behaviour that can be examined is purchases of market goods that are closely tied to the non-market goods of interest. The hedonic pricing method uses this approach, for example in relation to rents or property value. A second type of behaviour that might be examined is decisions made about where to go to spend one's free time, captured by studies that use the travel cost method. Revealed preference techniques make use of past behaviour, while stated preference techniques survey what individual people say they would do, when presented with alternative visions of the future. The contingent valuation method (and its variants) then infers value from that hypothetical behaviour.

In their book, Valuing Cultural Heritage, Navrud and Ready (2002) review each of these techniques and present the basic strategy motivating each, along with a discussion of the practical difficulties involved, especially with regard to valuation of cultural heritage goods. In general, the main studies have utilised stated preference techniques, especially contingent valuation, which uses survey methods to establish willingness-to-pay. The following section is based on Navrud and Ready's review.

7.5 Application of CVM to Cultural Heritage Goods

Navrud and Ready reviewed as many of the studies that valued cultural heritage as they could find upto 2002, most of which used a stated preference technique. As they stated, economists usually prefer to rely on actual consumer behaviour in response to economic incentives, as opposed to hypothetical statements about behaviour, which are believed to give more reliable indicators of value since mistakes result in real costs. Unfortunately they were unable to find many situations in which observed behaviour revealed preference for cultural heritage goods. Stated preference techniques, therefore, are often the only practicable approach to generating value estimates for cultural heritage goods, especially where the goods is a change in the characteristics of a building or monument.

There are good reasons for this since, in order to apply the revealed preference techniques, an observable decision must exist that reveals the individual's value for the goods (such as paying extra rent to live in a historic quarter or payments made to travel to visit a monument). For a number of reasons, these situations are rare for cultural heritage goods. It is difficult, for example, to disentangle the heritage value component of increased rents from other benefits of living in the same area. The contingent valuation method is much more flexible with regard to the goods that can be valued, as long as a credible scenario can be devised that allows the person being questioned to imagine the two situations (e.g. before and after cleaning of a building). This does not even have to be technically realistic – a non-existent conservation technique, for example, might be suggested as the programme that will bring about a cleaner building, and the responder would be shown artificial "before" and "after" images.

There is another important question, which relates to sample selection and the population it represents. The normal effects of using questionnaires, whereby people make assumptions about the importance of an object of enquiry, simply because somebody is asking them about it and they might frame answers to please the interviewer can generate an overfocusing effect within respondents. Caution must therefore be exercised when using these methods for valuing goods that are not widely known to the respondents, for example a cathedral that is being studied, since a respondent who had never heard about it prior to the interview may think that if the goods in question are so important that it is worth doing interviews about, then he must have some value for it. The effect is magnified if the derived values are scaled up to very large populations, which might result in very large aggregated estimates of WTP.

Most people accept the idea of public provision of cultural heritage, whereas other scenarios such as the clean up of a polluted river would be seen as the responsibility of the polluters. Contingent valuation does not work well in this type of situation and a contingent valuation scenario involving payment by individuals is often rejected by respondents. The reported experience of most studies involving cultural heritage goods is that the general public sees it as natural that the costs of provision will be borne widely.

Contingent valuation is the only valuation method that can capture non-use values, which are an interesting additional benefit that heritage can be clearly shown to have. The discussion above presented a number of such values – existence value, altruistic value, bequest value and option value. It is clear that for some objects non-use value will be very important (a simple example that immediately demonstrates the point might be the Taj Mahal, which many people would want to see preserved both for future generations or the general good, even if they never planned to see it in person.).

Valuing heritage programmes is very complex, since actions may prompt different responses (even in that same individual). An example might be cleaning a cathedral, which could bring out detail that was previously covered by soot but could look too "new," reducing its aesthetic value. Art historians argue that conservation or, especially, restoration may damage authenticity but these actions may also prevent complete loss of the object. Contingent valuation of goods with both positive and negative aspects is thus extremely challenging.

7.6 The Effects on the Local Economy

Restoration of a cultural heritage site is undertaken for many reasons but undertaking it usually also stimulates the local economy and employment and is likely to have a long-term positive impact on tourism. Therefore it is beneficial to consider this alongside the cost benefit equation relating to the site itself. The likely spin off effects of restoration projects occur in the construction work (materials and workers), ongoing maintenance expenditures and effects on tourism.

The tourism effect depends on the attractiveness of the cultural heritage, which may be of local, national or international interest. This is important in identifying the extent of the balance between local and/or regional economic impact.

The effects of a restoration investment include secondary and subsequent rounds of money circulating within the economy, made up, at least in part, of the salaries paid in the tourist sector and subsequent spending by tourism sector employees which will also benefit the local economy in part. In construction and maintenance phases too, a part of the investment goes to salaries in local restoration firms and locally sourced materials with secondary and subsequent effects of this expenditure from the spending of employees in local firms and the supply chain.

Multiplier values may be determined to estimate the economic impact of restoration and tourism-for-demand side effects of a restoration project, analysed with respect to the impacts of enhanced tourism activities and for supply side effects, related mainly to the restoration work. The overall analysis required examines local economic impact, mainly in local and regional towns, which can take the following forms.

- Demand side analysis based on visitor spend attributable to renovation
- Supply side analysis based on investment in renovation works.
- Income and employment effect.

A basic assumption is that any investment in cultural heritage with local tourist potential would be expected to contribute less to the economy than investing the same amount in a cultural heritage of regional or national tourist potential, since the visitor numbers are likely to be much smaller. It follows from this that the multiplier value should be higher, the greater the tourist potential of the underlying cultural heritage.

The economic context has an effect on the local economic impact of heritage restoration. If the economy is isolated (closed) but developed, a largely local work force and materials are likely to be used in the restoration process, whereas the workforce and/or materials involved in the restoration of a more urban and open area are often garnered from a wide area in the surrounding economy, thus reducing the overall impact on the local economy. This can be illustrated by the requirement for specialist skills such as where an ancient building needs special cleaning that is offered only by a small number of firms in the national economy for example. Small, local economies may lack specialist skills/firms and so a restoration project that required anything of a specialist nature might result in all of the investment being spent outside of the area.

The demand side economic impact of a cultural heritage restoration will be limited to tourism. The demand side effects can be broken down into additional visitors and consequently additional tourist expenditure arising after a restoration project is completed. The effects of increased tourism activity will be linked to the spending of tourists in the local economy – such as for accommodation and food.

7.6.1 The Multiplier Effect

The aim of multipliers is to determine the economic impact of an initial payment either on investment (from restoration expenditure) or increase in final demand (from increased tourism) by converting spending to income and jobs and capturing secondary impacts of tourist spending. The income to the region is reflected in the wages, salaries, rents and profits generated by tourist spending and also the revenues to local government units from taxes. The value added is the income and indirect business taxes generated by tourism and restoration.

The multiplier effect measures the total income effect after an initial payment and circulation of this payment in the economy. Payments used for imports etc. (see below) are termed leakage as the money leaves the local economy whereas payments used for salaries are likely to be respent within the local economy. Second round money circulation also leaks money as some will be saved up and others will be used for imports. Some will go to salaries and local goods and a third and fourth money circulation can begin.

Aggregation of the leakage from the money circulation relative to the initial payments gives the reciprocal value of the multiplier, such that:

Multiplier $=$ 1/proportion of leakage

A multiplier effect as described does not take the substitution effect into account, and represents only the local impacts. It is quite possible that the local economic benefits in economy A (through investment from economy B), are equal to the loss of economic activity in economy B through the reduction of expenditure in economy B to fuel the investment in economy A.

7.6.2 Leakage

Tourist spending can yield a distorted picture of tourism's impacts, particularly when tourists are buying imported goods. Leakage indicates amounts spent outside area and not recirculated locally. Leakage can be controlled to a certain limit depending on the nature of tourism planning. However other areas are outside of the influence of local authorities. The following are common examples of tourism leakage (McGahey, 1996)-

- Imports to pay for goods and services required or preferred by international tourists
- Commission to travel agencies or tour operators whose business is located outside the destination
- Franchise royalties and management fees paid to foreign hotel companies
- Profits to foreign workers and absentee landlords
- Interest paid on foreign loans
- Foreign exchange cost of capital investment in tourist facilities
- Advertising, promotion and publicity abroad
- Tourist use of credit cards and travellers checks that are drawn on foreign banks
- Wage remittances and savings sent home by foreign workers
- Overseas training of tourism personnel
- Savings of employees of tourism and tourism related business
- Taxes paid to higher government

7.6.3 Economic Impact of Tourism

An approach to estimating economic impacts of tourism (demand effect) is to directly survey tourists to estimate their spending. Estimates of spending can be translated into the resulting jobs and income in a given area using appropriate economic ratios (labour costs, and the share of labour costs of the total costs to the sector) and multipliers. The direct survey method is more applicable to estimating impacts of particular actions on a local economy. These more focused impact studies frequently also include multiplier effects of tourist spending on a region.

The basic equations are:

Tourist spending = Number of Visitors × Average spending per visitor

Economic impact = Number of Visitors × Average spending per visitor × Multiplier

Economic impacts may be estimated in terms of spending, sales, income, value added, tax revenues and employment. Estimating the number of visitors requires a clear definition of what a visitor (tourist) is and by what units tourism activity is measured in (e.g. person trips, person nights). Tourists are generally visitors from outside the region of interest (apart from heritage sites of purely local interest). Distinguishing between visitors staying in the area and day visitors is critical to the assessment of the impact of tourism; with staying visitors being more valuable (they will spend on accommodation and meals and may be expected to be in the area for longer periods).

7.6.4 The Employment Impacts of Cultural Heritage Restoration

The employment impact of tourism is linked to income and the multiplier. The tourism sector is characterised as labour intensive, offering jobs such as tour operators, travel agencies, hotel restaurant staff etc. The job contribution from tourism is often seasonable and part time – and therefore additional employment is often measured in "full time (job) equivalents" or "fte". Contribution to employment is more job hours than actual full-time employment. However, in local societies, especially if isolated or less densely populated, tourist earnings may generate an important additional income to society.

7.7 Case Study: Cost Benefit Analysis of Damage to Heritage Materials in Europe by Air Pollution

The Clean Air for Europe (CAFE) Programme (CAFE, 2008) is an initiative of the European Commission, introduced in 2001 with the aim of providing long-term, strategic and integrated policy advice as a means of reducing the adverse effects of air pollution on human health and the environment, including cultural heritage materials. There are several elements to the CAFE analysis, including predicted future emission levels, dispersion modelling centred around use of the EMEP model, integrated assessment modelling using the RAINS model and a detailed cost-benefit analysis using an updated and revised version of the ALPHA model. The cost-benefit analysis takes explicit account of damage to materials. It is intended to go beyond earlier work that was restricted to buildings that were not of significant cultural merit, though a full economic assessment of cultural heritage is still not possible.

7.7.1 Methodology for Corrosion Effects

Assessment of corrosion damage to building materials in everyday applications uses the following framework:

- Describe the stock-at-risk in terms of the exposed area of sensitive materials on buildings and other structures within each country (see Chapter 6 for further details);
- Describe background pollution levels;
- Describe climatic variables relevant to the response functions;
- Apply material-specific dose-response functions that provide estimates of the rate of corrosion, etc. (see Chapter 3 for further details);

- Assess the rate of deterioration against a critical loss of material (a level at which it is economically sensible to repair, clean, or otherwise maintain) to define change in maintenance or replacement frequency;
- Value using standard maintenance cost data from reference sources used by builders and architects.

Valuation of the loss of material requires some assumption to be made about human behaviour in relation to maintenance. In past work (e.g. ExternE, 1999; Holland et al., 1999) it has been assumed that the owners of buildings and other structures are perfectly rational and undertake maintenance once a "critical thickness" of material has been lost, and prior to secondary damage (e.g. wood rot in window frames) becomes a problem. "Critical thicknesses" were derived from consideration of the thickness of galvanised and paint coatings and observations of the amount of stone lost as an average across a façade before repairs would be initiated. The situation for cultural heritage is, of course, much more complicated. The method recognised that estimating stock at risk for cultural heritage poses problems, which has been discussed in detail in Chapter 6. It is noted that "Conclusions on the precise approach to be used for dealing with cultural heritage in CAFE are still to be finalized". Keith Bull, as part of a peer review of the methodology (Krupnick et al., 2005) felt that "it is important to try to answer this question since cultural heritage damage is becoming increasingly important to the public; more and more throughout Europe attention is being paid to our cultural heritage and the public have an increasing awareness of damage inflicted and greater expectations for repairs/preservation". These concepts are discussed in Chapters 3 and 9.

7.7.2 Methodology for Soiling Effects of Particles

Soiling of buildings by particles is one of the most obvious signs of pollution, especially in urban areas. It impacts on both "utilitarian" and historic buildings and causes economic damage, since increased frequency of cleaning, washing, or repainting of soiled surfaces becomes a considerable economic cost and can reduce the useful life of the soiled material (Newby et al., 1991). A simplified approach is often used that quantifies soiling damage based on cleaning costs (in the absence of WTP data). However, as this approach does not include amenity costs, estimates will be lower than total damage costs. Rabl et al., (1999) went further and looked at total soiling costs (i.e. the sum of repair cost and amenity loss). The study showed that for a typical situation where the damage is repaired by cleaning, the amenity loss was equal to the cleaning cost (for a zero discount rate); thus the total damage costs are twice the cleaning costs. The study recommended the following function:

$$S_i = a \times P_i \times \Delta TSP_i \quad (\text{where } a = b \times 2)$$

S_i = Annual soiling damage at receptor location i.

P_i = Number of people in location i.

ΔTSP_i = Change in annual average TSP (Total Suspended Particles) $\mu g\,m^{-3}$.

    a = WTP per person per year to avoid soiling damage of 1 $\mu g\,m^{-3}$ particles.

    b = Cleaning costs per person per year from a concentration of 1 $\mu g\,m^{-3}$ of TSP, assumed to be 0.5, based on Parisian data.

This function allows a site-specific assessment, linking reductions in particle concentrations with population.

Total national costs have been estimated for the UK by the Interdepartmental Group on Costs and Benefits, IGCB, a group of government economists and other experts that provides economic analysis and advice relating to the development and achievement of the UK Air Quality Strategy (DEFRA, 2001). The main approach adopted by the IGCB in its Second Report was that developed by Rabl (1999). The IGCB estimated soiling costs (due to PM_{10}) for 1998 and 2010 at £337 million/year and £177 million/year respectively. This implies an estimated £160 million reduction in cost of damage from building soiling over the period 1998–2010 on a "Business as Usual" scenario. The benefits in terms of reduced building soiling damage from additional pollution control measures were also considered. The estimated figures show an annual benefit of £12 million in 2010 from the additional transport measures (particulate traps, introduction of 10 ppm sulphur diesel and retrofitting) and £41 million for the stationary measures, implying a total annual benefit of £53 million in 2010.

7.7.3 Costs and Benefits in Europe

The CAFE analysis estimated that the cost of material damage in Europe due to air pollution was 1.13 billion in 2000, falling to 0.74 billion in 2020 when emissions will fall due to new EU legislation (see Chapter 9) to be implemented (CAFE, 2008).

The methods for quantification of damage to materials within CAFE follow work carried out by ICP Materials, described in Chapter 3, amongst others (e.g. the ExternE project). In confirmation of the position described earlier in this book it is interesting to note that the CAFE methodology states "The "impact pathway" approach works well for those applications that are used in every day life. This could in theory be applied to cultural and historic buildings. However, in practice there is a lack of data at several points in the impact pathway with respect to the stock at risk and valuation. As a result, effects of air pollution on cultural heritage cannot be quantified and thus need to be addressed qualitatively through the extended CBA framework". This reinforces the point that CBA analysis may be useful even where data is sub-optimal. The qualitative approach is set out below:

The "extended CBA" approach used by CAFE attempts to provide contextual information on impacts for decision makers, so that they can understand them better and, where appropriate, factor in their own views on the importance of unquantified impacts to the CBA.

The extended CBA is based around the provision of a series of datasheets that decision makers and stakeholders can refer to when evaluating costs and benefits. These contain a significant amount of descriptive information (qualitative and quantitative), aimed at the development of a better understanding of the effects and their likely importance.

7.8 Development of Datasheets for the Extended CBA

The extended CBA datasheets will contain the following types of information:

- Description of the impact, including components of "total economic value"
- Discussion of related impacts
- Confidence in attribution of impact to a specific pollutant
- Information on the distribution of impact across Europe (is it a "European issue" or something to be considered at a more local level?)
- Information on importance in economic or other terms, where available (e.g. from results of willingness to pay case studies, past estimates of expenditure to deal with specific problems, etc.)

An assessment will be made in the extended CBA of the likely significance of each unquantified effect, using a three point scale:

★★★	Effect will be significant at the European level
★★	Effect may be significant at the European level
★	Effect unlikely to be significant at the European level but significant locally
No stars	Negligible

The current conclusion of the CAFE programme is that damage to cultural heritage warrants a ** rating. The overall outcomes of the analysis are summarised below (Table 7.2).

In the case of cultural heritage, further information will be required to improve the assessment

- Description of impact, including which materials are at risk and which pollutants are of concern.
- Components of economic value – preservation costs, repair costs, amenity costs, etc.
- Strength of association of effect with air pollution (in this case, high, given the knowledge gained through research under the ICP Materials).

Table 7.2 Outcome of the extended CBA analysis

Effect	Preliminary rating
Health	
Ozone: chronic effects on mortality and morbidity	★★
SO₂: chronic effects on morbidity	★
Direct effects of VOCs	★
Social impacts of air pollution on health	★★
Altruistic effects	★★
Materials	
Effects on cultural assets	★★
Crops	
Indirect air pollution effects on livestock	★
Visible injury following ozone exposure	★
Interactions between pollutants, with pests and pathogens, climate...	★★
Forests	
Effects of O₃, acidification and eutrophication	★★★
Freshwaters	
Acidification and loss of invertebrates, fish, etc.	★★★
Other ecosystems	
Effects of O₃, acidification and eutrophication on biodiversity	★★★
Visibility	
Change in amenity	★
Groundwater quality and supply of drinking water	
Effects of acidification	★

Source: Holland et al., CAFE (2008)

- Discussion of the relative roles of pollution and natural factors in determining the rate of decay.
- Distribution of impact, perhaps including maps showing exceedance of various estimates of acceptable damage rate.
- Reasons for failure to quantify damage.
- Review of literature on valuation of cultural heritage, especially where pollution effects have been considered.
- Assessment of likely economic importance at the European scale.
- Assessment of likely economic importance at national and local scales.
- Conclusion on overall star rating.

Future research needs to pay more attention to the development of inventories of stock at risk, not just for cultural heritage, but for materials in utilitarian applications also. A second priority concerns assessment of the approach used to extrapolate from observations on small samples exposed in the field, to whole buildings. Any work that seeks to validate the results of impact assessment in terms of damage rate or economic effects should also be encouraged.

Another area in which further research should be encouraged concerns the aggregation of data on the economic benefit of reducing pollution damage to cultural heritage. It may be possible to do this from consideration of the way that other environmental criteria are aggregated in economic assessment, in other words, without the need for specific new.

7.9 Summary

Weighing the balance between cost and benefit is a useful contribution to the evaluation of where to invest scarce resources into the preservation of our heritage. Resources are not infinite, despite the fact that many objects are held to be "priceless". The major problem of cost-benefit analysis is putting a monetary value on costs such as pollution or noise, or measuring in financial terms value that people place on a much loved or beautiful object. Various techniques have been presented to show how the pollution damage costs can be estimated by use of dose-response functions to calculate the amount of damage caused by a given level of pollution, the cost of fixing it and the stock of materials affected. The value placed on improvements to heritage by individuals can be estimated by a number of methods, the best of which are based on surveys of their stated willingness to pay for repair. Such methods do carry the possibility of bias and are hard to transfer between objects and between countries.

Ultimately the arguments for investment must be presented, normally to governments or government agencies, for a political decision. The role of the economic analysis in making all of the costs and benefits transparent, including evaluation of alternatives and in quantifying what it is reasonable to quantify is that it allows a superior basis for political decision making. The valuation of non-monetary costs and benefits may be helpful to those responsible for heritage protection by making the rewards of investment more tangible to those making decisions in an arena of competition for resources.

Acknowledgements The foundation for this chapter was the REACH Project, funded by the EU: Environment and Climate Programme under Topic 2.2.4. PROJECT No: ENV4-CT98-0708 (REACH). The major economic contributions to this project were made by Navrud and Ready on indirect costs (who have since published this work and more in their book, *Valuing Cultural Heritage. Applying environmental valuation techniques to historical buildings and monuments*, see below). The UK Building Research Establishment and their sub-contactor, Ecotec Ltd, examined the effects on the local area and the local economy. BRE, with ITAM in the Czech Republic also developed a great deal of the material on direct costs, building on earlier work by all partners in the REACH Project. We are grateful to Milos Drdácký in particular. We are also grateful to Mike Holland and Paul Watkiss, CAFE, for their contributions to this chapter based on their presentation to the MULTI-ASSESS Workshop, London 2004.

References

CAFE (2008) Details of the CAFE Project Cost Benefit Analysis: http://www.cafe-cba.org/
"CULT-STRAT Project" – Assessment of Air Pollution Effects on Cultural Heritage – Management Strategies 2004–2007 Contract number: SSPI-CT-2004-501609.
DEFRA (2001) An Economic Analysis to Inform the Review of the Air Quality Strategy Objectives for Particles A Second Report of the Interdepartmental Group on Costs and Benefits http://www.defra.gov.uk/environment/airquality/panels/igcb/research/index.htm
ExternE (1999) DGXII (JOULE Programme) Externalities of Energy, ExternE Project, Report Number 7, Methodology: Update 1998. Holland, M.R. and Forster, D. (eds.).

Holland, M.R., Forster, D. and King, K. (1999) Cost-Benefit Analysis for the Protocol to Abate Acidification, Eutrophication and Ground Level Ozone in Europe. Report Number: Air and Energy 133, Ministry of Housing, Spatial Planning and Environment (MVROM), Directorate Air and Energy, ipc 640, P.O. Box 30945, 2500 GX The Hague, The Netherlands.

Holland, M, Hunt, A., Hurley, F. Navrud, S. and Watkiss, P. (2005) Final Methodology Paper (Volume 1) for Service Contract for carrying out cost-benefit analysis of air quality related issues, in particular in the clean air for Europe (CAFE) programme. AEA Technology.

Krupnick, A., Ostro, B. and Bull, K (2005) Peer review of the Methodology of Cost-benefit analysis of the Clean air for Europe programme.

Navrud, S. and Ready R.C. (2002) Valuing Cultural Heritage. Applying environmental valuation techniques to historical buildings and monuments. Edward Elgar publishing, UK. 279 pp.

Newby, P.T., Mansfield, T.A. and Hamilton, R.S. (1991) Sources and Economic Implications of Building Soiling in Urban Areas, Sci. Total Env., 100, 347–366.

Rabl, A. (1999) 'Air Pollution and Buildings: An Estimation of Damage Costs in France,' Environmental Impact Assessment Review 19, 361–385.

Sources of Additional Information

A useful primer on the theory of economic valuation, with case studies focussing on the heritage sector is:

Valuing Cultural Heritage by Navrud, S. and Ready R.C. Edward Elgar. 2002.279 pp.

"CULT-STRAT Project" – Assessment of Air Pollution Effects on Cultural Heritage – Management Strategies 2004–2007 Contract number: SSPI-CT-2004-501609.

http://www.swereakimab.se/web/page.aspx?pageid = 8529

Chapter 8
Risk Assessment and Management Strategies at Local Level

Tim Yates, Miloš Drdácký, Stanislav Pospíšil, and Terje Grøntoft

8.1 Overview

This chapter looks at ways that the techniques and procedures described in the previous chapters can serve as tools for owners and managers including local authorities responsible for objects of cultural heritage. All of them have their benefits and usually a combined use of some of them can be an efficient tool in the efforts of reducing the risk and consequently the maintenance cost. The chapter examines ways that mapping and modelling can be used to build extra layers onto the earlier maps and extrapolations presented for pollution. Pollution maps are developed into corrosion and soiling maps by application of the dose-response functions presented in Chapters 3 and 4. This chapter returns to the theme of scale and shows how broad patterns may be transformed by local influences.

Most of the monitoring discussed in earlier chapters refers to national and regional scale monitoring of pollution and environmental parameters. Monitoring, of course, is also important at a much more local scale. This does not simply mean micro-scale measurement of these parameters but the monitoring can be extended to the evaluation of effects too. A simple toolkit to evaluate effects at a building or monument that integrates the concepts discussed in the preceding chapters through provision of suitable indicator materials for exposure in situ is described. This can be exposed for a year and then returned to the laboratory for analysis. Similar approaches can be used indoors too and these are briefly reviewed.

Important as monitoring is, it simply provides a snapshot of the situation current at the time measurements are made. As has been discussed in Chapter 2, models can be used to extrapolate measured values into a more general view or even to future scenarios. This is extremely useful in trying to decide what to do in a given situation as different policy interventions can be modelled and evaluated. Chapter 2 demonstrated that a model (the AirQuis system) could be used to reproduce observed PM_{10} concentrations rather well. The system was

T. Yates (✉)

BRE-Building Research Establishment, Ltd., Garston, Watford WD25 9XX, UK

e-mail: Yatest@bre.co.uk

J. Watt et al. (eds.), *The Effects of Air Pollution on Cultural Heritage*,

DOI 10.1007/978-0-387-84893-8_8, © Springer Science+Business Media, LLC 2009

consequently considered well suited for studies of air quality exceedances of the proposed limit values of the EC Daughter Directive. This chapter shows how the same modelling results for PM_{10} may be used to investigate soiling of building surfaces in Oslo. This type of example is a useful demonstration of the way that a spatial representation of the potential damage effects (or their reduction) may both show ways to optimise local interventions and also to highlight the worst affected areas to set priorities for remedial action.

Another modelling case study is presented to illustrate the way that the necessarily broad sweep of the regional and even city-scale models can be further refined to focus on individual buildings or components of them. This is an important scale for the majority of conservation actions, of course. The study discusses the way that pollution distribution may be influenced by the wind load on historic structures. Common experience suggests that there are complex micro-scale variations across structures that alter the general deposition patterns discussed up until now. Scientists undertake simulation studies to try to understand the processes that control this deposition. An experiment on two scaled mock-ups of typical church towers to permit an evaluation of pressure and flow patterns are shown as examples.

Cost-benefit analysis of environmental strategies related to cultural heritage objects was introduced in Chapter 7. It is not simple and is influenced by many side effects, local constraints and conditions and so only very approximate or general tendencies can be revealed and used as the basis for comparison. However, it is an important area and so it is necessary to try and make some progress in understanding the interaction of different factors and benefits – in both social and economic terms – than can result from planned conservation strategies. As shown above, studying the cost and benefits of conservation strategies and policies requires an understanding of the interaction of materials and their environment. Understanding of the deterioration of materials can then be used to assess the service life and potential time to failure of materials and components, but the cost of the "failure" or avoidance of the "failure" requires consideration of the possible interventions – maintenance, conservation and renovation. Each of these can be defined and the implications of them for the built heritage quantified. These theoretical studies need to be complemented with observations made on site – in terms of interventions, cost and the periods between different forms of intervention. So, once again, case studies are presented, from France and the Czech Republic, which illustrate the progress that can be made despite the complexity of the buildings and structures involved.

The chapter then changes scale once more, to examine an important question for many heritage objects – to what extent is it possible to draw inferences about pollution levels indoors from data and information on ambient levels outdoors? This chapter will extend the discussions to the impact of pollutants that originate in the external environment but which present a risk to museum collections and to the interior fixtures and fittings of the built heritage. It will present an overview of the types of damage that air pollution causes to indoor cultural heritage materials. It will also examine ways of minimising its impact on indoor air quality, as well as techniques for dealing with its effects on buildings – i.e. the degradation and bio-deterioration inside buildings. It will consider the various sources of air pollutants and the role display cases and museum buildings can play in the control of air pollutants. The concept of "tolerable" levels for air pollutants and material damage in internal environments and their basis will be discussed.

The discussion of the indoor environment thus ends with a discussion of policy development. In the more general area of conservation policy at a local level, there is often a perception that the interval between interventions is decreasing and the evidence available shows that this is the case at least in some locations. However, it is important to look at these interventions against the background of conservation principles being followed at the time of a particular intervention. The actions and changes that are deemed acceptable are often embodied in a "Conservation Philosophy" that contains views on the appropriateness of maintaining a "like for like" replacement of materials, whether some degree of "improvement" may be desirable (for example if the past use of a material or design has been unsuccessful should we repeat this "mistake" knowing that it will fail again at some point or should we look for a degree of improvement?), and on the "best practical options" including consideration of whether they can be achieved at no excessive cost. It seems clear that in the past, particularly the period from 1850 to 1950, that interventions tended to be far more radical in many countries, often involving the rebuilding of substantial parts of buildings or the replacement of areas of stonework and glass. In more recent times the desire to keep interventions to a minimum has resulted in less invasive, but more frequent, work.

It is possible to identify a hierarchy of current guidance relating to conservation and maintenance strategies and these in turn inform the practitioner of the techniques which are applicable in different circumstances. At the top of the hierarchy are Conservation Principles, often embodied within charters or policies, for example, the Venice Charter or the Stirling Charter. Following the guidance in these documents will generally result in a holistic approach. There is also an acknowledgement that whilst the historic environment must allow for new architecture and modern lifestyles there should be a presumption in favour of conservation. The charters also emphasise that conservation and restoration must have recourse to all the sciences and techniques which can contribute to the study and safeguarding of the architectural heritage. A representative sample of national and regional guidance and the legislative organisations established to undertake it will be presented.

8.2 Monitoring at Local Level

The important purpose of monitoring the exposure of materials and objects both in the outside environment and indoors in museums and archives to pollution is to assess the degrading influences of the pollutants. Objects and collections are made from different materials and combinations of materials and have different vulnerabilities to different pollutants. The dose-response function development for inorganic materials such as stone and metals may need to be extended to include organic materials such as polymeric paper, textiles and leather or molecular dyes. In reality, most natural and synthetic materials have been used to manufacture cultural artefacts and the list of expected degradation effects of pollutant – artefact combinations is long (see Tetreault, 2003; Blades, 2000). As discussed above, generally the pollutants have oxidising, acidic and/or soiling effects on the artefacts. Typical oxidation phenomena are corrosion of metals or photo-oxidation of organic polymers. Acids increase the potential for many corrosion processes and the solubility of calcareous materials, and are known to break bonds in polymeric cellulose. Particulates and dust soil, discolour and can wear surfaces and can take part in degrading chemical processes.

The range of processes that degrade the objects is large as is the range of pollution monitoring techniques and preventive conservation methods and strategies (Taylor et al., 2006). Environmental monitoring strategies used in preventive conservation can be divided into two categories: parameter monitoring and dosimetry (Blades et al., 2006).

8.2.1 Parameter Monitoring

The most common method of environmental monitoring has been parameter monitoring, where scientific measurements are made on numerical scales of relevant parameters such as temperature, relative humidity, and light and air pollution. What these data mean for preventive conservation is then interpreted using background knowledge from scientific studies of the interaction between materials and levels of the parameter, either through accelerated ageing tests (see e.g. Zinn et al., 1994) or natural ageing in field tests (e.g. Larsen, 1996). The latter method is much rarer than the former because of the long timescales of natural ageing and the difficulty in collecting historic data about exposure conditions throughout the lifetime of an object.

Background knowledge from these sources underpins the formulation of standards (e.g. CEN/TC346) and guidelines for preventive conservation (Taylor et al., 2006; Tetreault, 2003). However the data used are subject to many uncertainties such as those in extrapolating from accelerating ageing to what actually happens more slowly under ambient conditions. By contrast the methods used to measure environmental parameters are generally much more precise. It follows therefore that, at least in an early warning strategy; a semi-quantitative measure of environmental quality may well suffice. Therefore measurement by dosimetry may be entirely sufficient and has the advantages of often being easier and cheaper to carry out and often easier to interpret (Blades et al., 2006).

8.2.2 *Dosimetry*

Dosimetry can be thought of as the inverse of parameter monitoring: in parameter monitoring the potential for deterioration is inferred from environmental measurement. In dosimetry some form of sacrificial material that responds similarly to the materials of interest is exposed to the environment, and from its deterioration, the quality of the environment is inferred.

Some examples of dosimeters include the sensitive potash-lime-silicate glass dosimeter from the EU "AMECP" (EV5V-CT92-0144) (Fuchs et al., 1991; Leissner et al., 1992), the LightCheck devices developed as part of the EC "LIDO" project (EVK4-CT-2000-00016) (Bacci et al., 2003), the resin mastic covered quartz crystal microbalances from the EU "MIMIC" project (EVKV-CT-2000-00040) (Odlya et al., 2005), the organic polymer dosimeter from the EU MASTER project (Dahlin et al., 2005), the blue wool standards (Bullock and Saunders, 1999) and metal coupons of lead, copper and silver (Oddy, 1973). Ideally dosimeters are relatively easier to make, or cheaper to buy than most parameter monitoring techniques. On the simplest level their response is a visible change. They are often amenable to more detailed analysis, if needed. For instance, the corrosion layers on metal coupons can be subject to various spectroscopic analysis techniques, and the degree of light fading of a dye can either be compared with a card strip or quantified with a colour meter.

Other defining characteristics of dosimeters is that they respond in a synergistic way to the overall "aggressiveness" of the environment integrating the effects of all the different parameters present into a single response, and that they respond more quickly than collections material.

The generic response has advantages over parameter monitoring, where when we monitor an environment we assume we are measuring all the relevant parameters and may have to employ a range of techniques to do so. Some dosimeters respond greatly to one factor, e.g. light fading and for practical purposes can be considered as single parameter dosimeters, but will however also respond more subtly to other factors such as air pollution and temperature. For other dosimeters the responses are more evenly distributed. For instance, the corrosion of lead coupons requires organic acids and a sufficient degree of humidity both to be present. The reaction is probably further accelerated by temperature and the presence of other pollutants. The generic response is useful for a device that is intended to give an overall indication of environmental quality. It is less useful for diagnostic purposes in that where a problem has been found, there is no clear indication of which parameter is causing the problem. In this case more diagnostic monitoring techniques would need to be employed to identify the specific cause(s) of the problem.

Dosimeters need to respond more quickly than the collections material to give information earlier than obtained from examining the collections material itself. For example, in the case of light dosimeters the response can be speeded up by using very light-sensitive dyes that would not have any practical use as pigments but are useful for dosimetry. For other materials such as silver coupons, it is less obvious how the response can be speeded up compared with a silver object. In practice this can be done by making sure the surface is clean of any passivating oxide layers by scrubbing with an abrasive before exposure.

The response of a dosimeter can be related directly to the environment it is exposed in and indirectly to the material we wish to conserve, in that environment, by extrapolation. Extrapolation can be performed as part of a calibrated system, where the response from the dosimeter material is calibrated against an environmental quality hierarchy. This is the way it was developed in the EU-MASTER project. In the MASTER project it was calibrated against the generic building environments, supported by literature information (see e.g. Dahlin, 2006; Tétrault, 2003; Sebera, 1994) on the deterioration effects of environmental parameters on materials.

Monitoring of environmental parameters and pollutant concentrations in indoor environments in conjunction with the application of indoor air quality models (discussed later in the chapter) can provide useful information on the preservation conditions for materials inside museums and historical archives. Degradation of materials is a long time process. Therefore the monitoring and modelling of average concentration values over long time periods is of importance in assessing effects on cultural heritage objects and in determining the adequate preventive conservation strategy.

8.3 In Situ Evaluation of Effects – Toolkit for Damage Assessment

People in heritage management are often interested in the level of corrosion attack on specific locations at their object of interest. To respond to this demand the MULTI-ASSESS project developed a Rapid Tool Kit. The Rapid Tool Kit was designed with two versions: *Basic* and *Complete*. Both versions contain a rack with reference specimens for the assessment of corrosion and soiling of materials or material groups of interest. The reference materials were chosen as material indicators representing metals (carbon steel and zinc) and porous stone materials (Portland limestone). They were selected because they had been used in many exposure programmes and have reliable dose-response functions and sufficient degradation after one year of exposure. The *Complete* version also includes passive samplers of the relevant pollution parameters. The exposure period for both kits is one year in total to even out the effect of seasonal variation and to get reliable data for comparison with the tolerable corrosion levels.

In the following, the two versions of the kit will be described more in detail with reference to Fig. 8.1

The *Basic* kit includes:

- 1 exposure rack
- 3 test specimens of carbon steel to be exposed at 45° angle facing south. The carbon steel specimen has a dimension 75×100 mm with a thickness of 1 mm and follow the recommendation given in ISO 9226 and ISO 11844-2. The

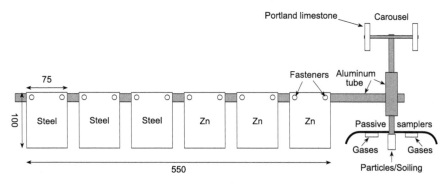

Fig. 8.1 Rack for the rapid tool kit – complete

carbon steel follows ISO 3574, CR 1: max 0,15% C, max 0,04% P, max 0,05% S, max 0,6% Mn

- 3 test specimens of zinc (> 99.9%) 75×100 mm with a thickness of 1 mm to be exposed at 45° angle facing south.
- 3 test specimens of Portland limestone mounted vertically in three different directions on top of the rack. The Portland limestone is of unified quality and has a dimension of 50×50×10 mm with a centre hole for the attachment.
- It is also possible to expose 3 samples of one type of local stone on the rack (optional).
- 2 test specimens of passive samplers for soiling measurements mounted sheltered from rain. Material for soiling is a Teflon filter paper attached vertically to a plastic rod.

The *Complete* kit includes all the elements of the *Basic* kit and in addition:

- Metal rain shields that are attached to the rack (ref. Fig. 2.6)
- bi-monthly passive samplers for the pollutants SO_2, NO_2, HNO_3 and O_3
- bi-monthly passive samplers for measurements of particulate deposition (mass and chemical speciation)

The passive samplers are placed under the metal rain shields protected from the rain and with the filter side facing down. The climate parameters (temperature and relative humidity) and rain (amount, pH) that are needed can be obtained from local meteorological sites.

Data from the tool kits will show if tolerable pollution levels are exceeded and the way that the environment is affecting deterioration and soiling of the relevant materials. If the exposures are repeated, for example every third year, the results will give trends of pollution and deterioration, which can be used as control of the effectiveness of strategies and measures for reduction of pollution and as an early warning of unexpected increases of corrosion due to new sources or combinations of pollutants.

As an example, monuments included on the UNESCO World Heritage List have special requirements for monitoring. Organisations and individuals being responsible for these monuments can use the kit to assess if their site is properly maintained and also give indications on possible plans of action, if needed, in the regular reporting to the UNESCO Heritage Fund Committee.

8.4 Mapping and Modelling Pollution and Risk of Corrosion and Soiling

As discussed above, monitoring on its own is not ideal for developing policy or evaluating its success. Running scenarios related to specified policy options allows a manager to choose between different potential actions and to set priorities for action.

Applications of the AirQuis model to studies of air quality exceedances of the proposed limit values of the EC Daughter Directive were shown in Chapter 2. As shown in Chapter 4, particulate matter is the main cause of soiling of buildings and monuments. Although the main driver for reducing the levels of particles, in Oslo as elsewhere, is concern about health aspects, the methodology and model calculations may nonetheless be used for studying soiling, as the following case study shows.

8.4.1 Case Study – Model Calculations and Scenario Modelling of Urban Levels of Particulate Matter and Risk of Soiling of Buildings in Oslo

The study was started with a reference simulation for the year 2003. The PM_{10} results from the 2003 simulation were compared with available measurements within the cities to make sure that the model system worked properly. Moreover, some adjustments of the source strength of the coarse fraction particles, i.e. $PM_{10} - PM_{2.5}$, were also implemented as part of this verification/validation work. Second, the emissions were altered to reflect expected and already adopted regulations in technology and fuels in the period up to 2010. Based on this updated emission inventory, a Reference 2010 simulation was performed. This simulation included exposure-soiling calculations in order to indicate the exceedance level to be expected if no abatement measures were implemented. Third, a set of abatement measures was defined. By incorporating the effect of these measures on the Reference 2010 emission inventory, a number of Scenario simulations were performed.

Ambient air concentrations and expected soiling of building surfaces were calculated both within a domain-covering, two-dimensional grid with a quadratic $1\,km^2$ grid size (hereafter simply referred to as grid values) and in the positions of buildings located close to the main road network (hereafter referred to as building point values).

Figure 8.2 shows the result for the reference simulation for 2003.

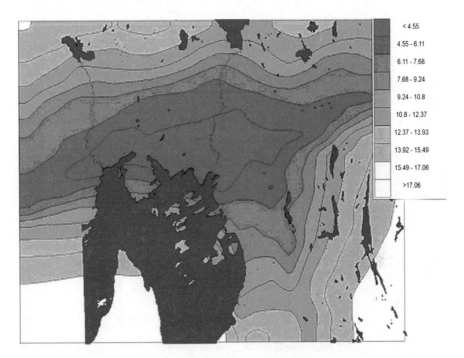

Fig. 8.2 Cleaning interval in years at 35% loss of reflectance calculated for Oslo for 2003 PM10 values using the MULTI-ASSESS dose-response function for soiling of limestone

Figures 8.3 and 8.4 show results for the reference scenario for 2010 with modelling of yearly averages in grid cells of 1 × 1 km and in building points along roads respectively.

Figures 8.5 and 8.6 show results for the least invasive abatement scenario for 2010, with a 10% reduction in the use of studded tyres, from 85 to 95%, with modelling of yearly averages in grid cells of 1 × 1 km and in building points along roads respectively.

The maps show cleaning interval in years at 35% loss of reflectance on limestone. The target value proposed in the MULTI-ASSESS project (Kucera, 2005) and to the CAFE (Clean Air for Europe) initiative of the European Environmental Agency is cleaning every 10–15 years. For corrosion processes a tolerable n-value of 2.5 times exceedance of the background value was proposed. Similar mapping of absolute degradation rate or exceedance, n-values, can be performed for corrosion and other degradation effects.

Comparison of the maps in Figs. 8.2, 8.3, 8.4, 8.5, 8.6 clearly shows that a reduction in PM_{10} and consequently longer cleaning intervals are expected in Oslo in 2010 as compared to 2003. In 2003 cleaning should have taken place at

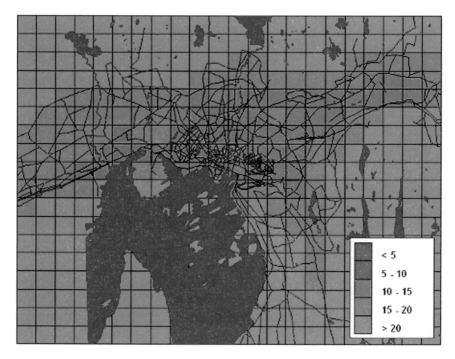

Fig. 8.3 Oslo Reference 2010: 85% non-studded vehicles + scaled emissions from wood burning. The colour scale gives the cleaning interval in years at 35% loss of reflectance on limestone

frequencies of 5–10 years for most of the city centre (Fig. 8.2) whereas the recommended cleaning interval for the city centre would be 10–15 years in 2010 (Fig. 8.3). With the relatively mild, and easy to implement, abatement measure of increasing the amount of non-studded tyres from 85 to 95%, parts of the city centre is expected to need cleaning less frequently, every 15–20 years (Fig. 8.5) which is proposed to be tolerable. However a comparison of Figs. 8.3 and 8.4 shows that the situation is less good for a number of buildings along trafficked roads both in the city centre, along the ring roads and on the main roads connecting the city to the north east. Here cleaning would be needed every 5–10 years for a large number of buildings and even more frequently than every fifth year for a few buildings, in 2010. The reduction of the use of studded tyres by 10% reduces the number of building points with exceedances significantly so that only very few buildings in the city centre and markedly fewer buildings along the ring and connecting roads would be in this category (Fig. 8.6). Still a few buildings in the most trafficked hot spots would need cleaning even more frequently than every fifth year. Previous modelling of corrosion effects in Oslo for 2003 did not show exceedances of tolerable levels. For Oslo the soiling effect

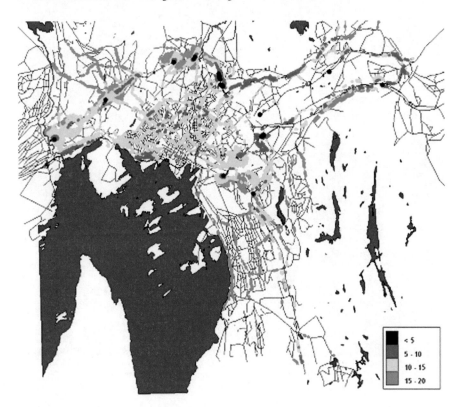

Fig. 8.4 Oslo Reference 2010: Figure 8.3 but with results for building points along road links

seems to be more serious than corrosion in terms of exceedance of proposed tolerable levels and the estimated costs of cleaning.

Figures 8.4 and 8.6 show all building points along roads where increased concentrations are expected compared to the urban background. Maps of the stock at risk of cultural heritage buildings could be superimposed on the maps in Figs. 8.2, 8.3, 8.4, 8.5, 8.6 to assess the soiling of those buildings.

The methodology of abatement – dispersion modelling, described above, is well suited for the study of degradation effects of air pollutants on the stock at risk of cultural heritage in urban areas. Models with different resolution exist for different European cities. The research institutions administering the models can run them on demand. Abatement – dispersion modelling can inform environmental authorities about the added utility of reduction of pollutants in terms of reduction of material degradation generally, and more specifically about possible abatement strategies to reduce PM_{10} to levels regarded as tolerable in terms of cleaning intervals and soiling costs. The next case study looks at an even smaller scale of modelling study – single buildings.

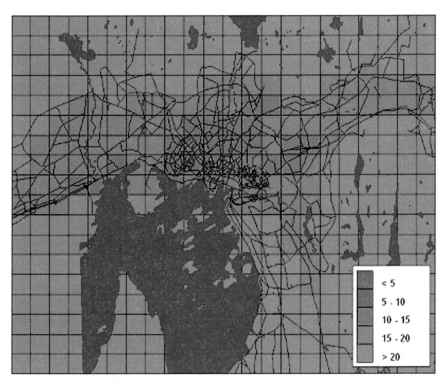

Fig. 8.5 Oslo 2010 Scenario 2: As reference 2010, Fig. 8.3, but with 95% non-studded vehicles

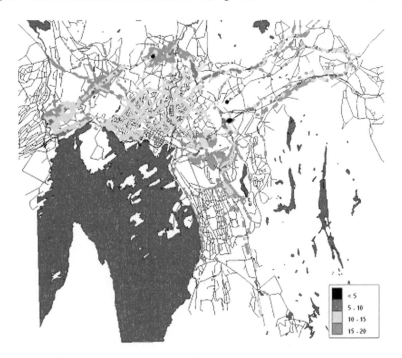

Fig. 8.6 Oslo 2010 Scenario 2: As reference 2010, Fig. 8.4 but with 95% non-studded vehicles

8.5 Case Study - Air Movements Around Complex Structures, Including Cultural Heritage

The processes of deposition and damage discussed in Chapters 1, 2, 3, 4 are necessarily generalised patterns over a number of different scales – they represent what is likely to happen to given materials, on average, within a region with a given level of pollution. But, of course, it is well known that superimposed on these patterns of exposure and depositions are many much smaller scale variations. All buildings and structures are subjected to variations in wind loads, for example. This in itself can be a threat to the structure – for many buildings, wind loading is the dominant design factor. Wind damage to buildings and structures is a severe problem throughout the whole of Europe. For example, during 1990, there were eight major windstorms, which caused widespread damage in at least 10 countries. The total loss from the windstorms in one year was estimated at 20,000 million Euros. It is not just in severe wind storms that building damage occurs, every year in Europe considerable numbers of buildings and structures are damaged from even relatively moderate wind storms. The high levels of wind damage place a severe financial burden not only on national economies in general but also on businesses and the public who have to pay for the repairs.

Wind primarily causes loading and mechanical damage of the structures; nevertheless, it also increases or decreases the chemical action of water and gases on cultural heritage objects. Flow around structures substantially influences the deposition of pollutants, biological corrosion, cycles of drying and wetting as well as mechanical wear of the attacked surfaces. Wind transports water, salts, dust and gases to the object or can conduct them away. It is clear then, that air flow effects on open-air located historical structures or sculptures can be quite disparate. Because of the synergistic effects of wind mentioned above, the repair and maintenance cost may be especially high in the case of historical structures. Some studies show that older structures such as roofs, and especially roof tiles, are more susceptible to damage than newer ones – perhaps at even twice the level. The roof failures may be partly due to ageing which make tiled roofs more sensitive to storms, while the more balanced distribution of roof damage shown by other data are due to exceptional wind velocities which induced stress levels closer to the ultimate limit state. As Fig. 8.7 shows, historical objects are complex and therefore vulnerable to the type of effects described.

Study of wind effects and an effective design of mitigation measures are dependent on a suitable ranking of structures and elements vulnerable to wind effects and models that evaluate the effects. Analysis of the flow around a building and identification of most dangerous places of the structure is generally very demanding task even for relatively simple geometrical shape. The numerical solution depends on many parameters, among the most significant of which is the wind velocity vector and dimensions of the body. This case study presents the results of numerical model validation using a wind tunnel

Fig. 8.7 A typical gothic tower, with sharp edges used as the prototype tower for numerical studies

benchmark replica of a typical historical tower. Figure 8.8 shows examples of some simulations of flow using complex mathematical models in the computer. Understanding flows in this way helps us to see where strains occur and where variations in pollution patterns may occur.

The model shows significant changes on the streamlines pattern and pressure distribution. The sharp boundary between dark and light areas, i.e. high and low speed on the left graph of Fig. 8.8 disappears and there is a smooth transition zone instead (right). Such a boundary would represent very high pressure differences in the free field between two adjacent points which is unrealistic. Moreover, in the vicinity of the small towers and close to the spike, the velocity vectors do not change the direction, which does not correspond to

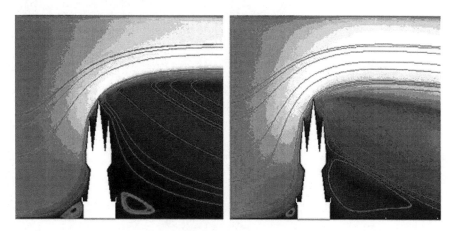

Fig. 8.8 Modelled velocity streamlines for a simulated tower

the real situation in the wind tunnel, where we observe genesis of eddies and vortices. This has been also observed during the flow visualization around the model of a tower.

Another modelled example can be seen in Figs. 8.9 and 8.10, which show the analysis of the wind load distribution on the historical tower roof and its changes due to different configuration of the roof geometry depending on the presence or absence of turrets.

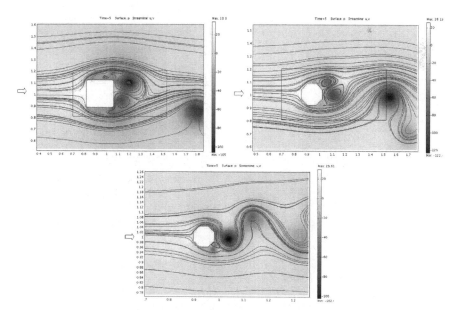

Fig. 8.9 Streamlines characterizing the flow around the top of an old historical tower without spires. Dark colour represents lower pressure

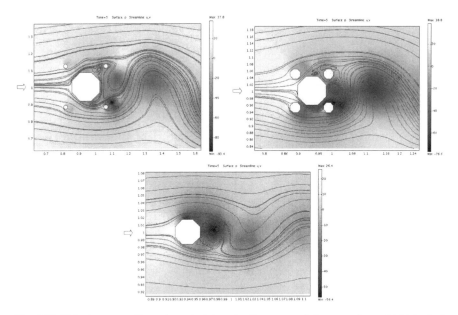

Fig. 8.10 Velocity streamlines characterizing the flow around the top of an old historical tower with small towers

The results are stationary and periodic and represent pressures and stream-lines distribution around the tower roof in certain arbitrary time and in certain elevation. The vortices and higher wind speed domains create differences in pressures and therefore zones of higher wind loads around the top of the tower.

The top left figure is the top of the tower, the right figure shows the flow around the middle part of the tower and the bottom graph represents the pressure and velocities distribution at the lower part of the tower.

The presence of the turrets obviously influences the wind load distribution. Generally speaking, they cause zones with very high pressure and suction, but in some cases, their presence leads to very low pressure in some zones, because the so-called backward flow influences the pressures.

This example demonstrates the usefulness of a numerical approach from the qualitative point of view and moreover the numerical recipes, suitable for certain physical phenomenon, also give reliable quantitative results.

8.5.1 Wind Tunnel Tests

The measurement of wind forces on historic buildings differs from the common test on modern structures. Even slight deteriorations caused by the wind may be more costly than on ordinary buildings due to the historic value of the fabric.

The next example brings together wind tunnel measurements made in Prague BLWT (Boundary Layer Wind Tunnel) with full scale measurements, both historical and current measurements, and modelling calculations. This results in a unique data-set which provides a basic understanding of the actual mechanism of drift built up in combination with strong and moderate winds. These data are used to calibrate the statistical models and facilitate their extension to a much wider range of typical detail and building shapes.

Several typical shapes for the historic roofs and structures have been selected, mostly for old town towers. Though variable in many details, they have been divided into two groups according to the aerodynamic shape of particular details of interest (i.e. pinnacles, cupolas, etc.) that very often characterise two major architectural styles; Gothic and Baroque.

Computer models of the flow around those typical tower structures including particular and most exposed details have been developed in two- and three-dimensional space. Currently these are based on assumptions of stationary flow with very low turbulence intensity.

8.5.2 Modelling of Air Flow Around Towers

The aim of the wind tunnel measurements and full-scale measurement of a complicated and often unique building is to help mostly in calibrating the Computational Wind Engineering (CWE) models. The measurement system consisted of 16 sensitive low-pressure sensors. They read analogously the pressure through the system of small tubes and specially bored miniature holes on the surface of the model. The scale of the models is naturally very important. The scale 1:20–1:30 has been selected as it complies with geometrical requirements of the testing facilities and do not interfere with similarity laws.

It is not possible to be certain that the flow patterns are completely realistic, a common problem in the flow field analysis or aerodynamics. To improve the understanding of the mathematical model, the study prepared the measurement of the pressures at particular points of the structure (roof, small towers). The inspiration for the model design was taken from the St. Týn church from the Old Town Square in Prague as there are plans towards the wind speed measurement in situ on this tower. The scale of the model is 1:25. The photograph of the model can be seen in Fig. 8.11.

Two balsa-wood models of prototype have been equipped with 16 pressure transducers, connected to the roof surface by optimized Teflon tubes (see Fig. 8.13 and 8.14). The pressures serve as the correction output for the

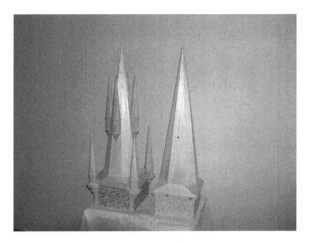

Fig. 8.11 Two balsa models of a typical (neo) gothic or tower roof

numerical studies. The pressure taps are schematically shown in the Fig. 8.12. Even with limited number of transducers, a "complete" pressure field at the roof surface can be obtained by means of gradual rotation of the model by about 45°.

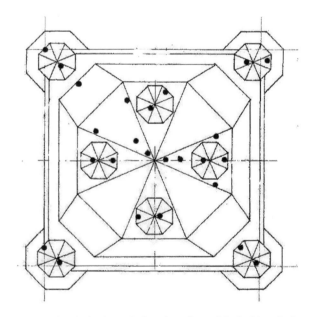

Fig. 8.12 Pressure taps (black dots) on the benchmark model of a historical tower

Fig. 8.13 Miniature hole in the model roof, on the removable turret

Fig. 8.14 The tube system connecting the pressure measurement spots with pressure transducers inside the model

8.5.2.1 Results and Application

The experimentally obtained pressure time histories, namely the mean and RMS values are calculated and compared with numerical calculation. The experimental and numerical mean values are shown in Figs. 8.15, 8.16, 8.17, and 8.18 for illustration. The results are very reliable. This does not mean complete agreement, but the fact that the analyst may identify the weak structural points, if applied properly, by a relatively cheap numerical procedure. The white arrow stands for the wind direction. Fig. 8.19 shows the flow direction using a smoke trailer.

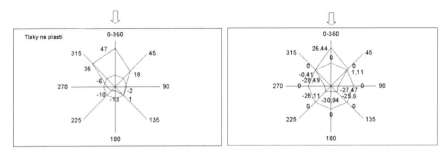

Fig. 8.15 Comparison of experimental mean pressures values (*left*) and numerical results (*right*) on the top of the simple (turrets-free) tower

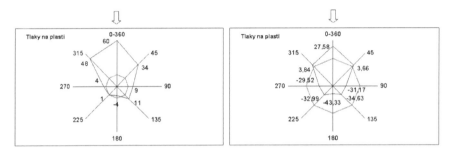

Fig. 8.16 Comparison of experimental mean pressures values (*left*) and numerical results (*right*) ad the mid cross-section of the simple (turrets-free) tower

Fig. 8.17 Comparison of experimental mean pressures values (*left*) and numerical results (*right*) at the upper part cross-section of the tower with turrets

The numerical procedure, if validated in some reliable extent, may be used further on for analysis of details like it is depicted in Fig. 8.20. This shows degradation of a structural element by the synergistic wind action and it is one typical example of usage of the method on a local scale. On a global scale, the procedure may help an expert in distinguishing the areas of the highest negative soiling exposition as it is presented for example in Fig. 8.21.

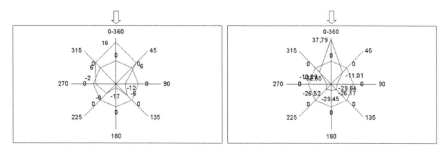

Fig. 8.18 Comparison of experimental mean pressures values (*left*) and numerical results (*right*) at the middle cross-section of the tower with turrets influencing the flow

Fig. 8.19 The view of the scaled (1:25) validation model of the tower in the wind tunnel during simultaneous pressure measurement and flow visualization using smoke trailer

Fig. 8.20 Stone deterioration of the Black Tower in Ceske Budejovice by synergistic wind action

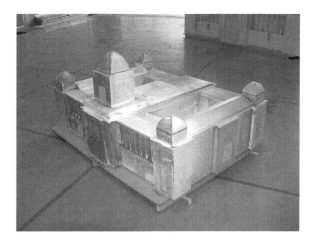

8.6 Practical Life Cycle or Life Time Estimates as a Base for Management Strategies

Air pollution damage to architectural heritage is a problem that can be reduced by establishing a local air quality policy and a conservation strategy that includes a heritage management policy. In reality, the decision about improvements of local air quality in historic settlements is seldom influenced by conscious evaluation of environmental effects on cultural heritage, though there are examples of managing traffic to reduce soiling of buildings (such as the EMIT Project in Oxford). Such effects are usually not quantified, and even if the air pollution is monitored, the measured values will usually exceed the thresholds for possible harm to living organisms before the effects on the built environment are considered. On the other hand, the development of a heritage management policy is more influenced by knowledge of the local situation and conditions of cultural heritage in cities including the environmental effects.

As discussed in considerable detail earlier in this book, diagnostics and monitoring of environmental damage has focused on research of material loss in exterior exposure and the results supported derivation of dose-response or damage functions which can be used for theoretical life time estimates for selected materials. However, cultural heritage objects – with the exception of some sculptures – are quite complex assemblies made of different materials and there are no suitable theoretical models for life expectancy computations. Here the weakest links might be analyzed or the object assessed as a whole. The latter approach has been advantageously applied in assessing soiling of façades.

Generally speaking, the time period between interventions for given historic materials and systems can be calculated using the dose-response functions and an agreed limit of tolerable corrosion before action, or it can be estimated from an evaluation of practical experience. The second approach inherently takes into account the local situation and any increase in air pollution due to local sources within a limited radius of possible impact. Of course, reliable long-term monitoring of both environmental and maintenance or construction work data is a necessary condition for assessing the typical time period between two interventions.

Such data is usually not available, and we have to use archive and/or comparative studies and interpolations or extrapolations for life time predictions. The lifetime is defined as a mean time to failure (MTTF) or a return period (RP), which is dependent on the data available. The MTTF is estimated from the known lifetimes of several similar objects (or elements), the RP is calculated from data known for one object on which an identical element has failed several times during the life of the object. The practical life time value, changes in response to the environmental factors, conservation technologies, materials and state policy as well as to changes in critical damage or serviceability definitions. Examples of return periods are given in Fig. 8.22.

The importance of knowledge of the intervention history of architectural heritage is obvious, and so there is an urgent need to start a wide-scale monitoring of local environmental data and to relate the data to repair and maintenance work on individual objects. The most favourable effects are in cities, where local policies can be created and put into practice with a realistic possibility to monitor the benefits and review the cost efficiency of the measures. Such monitoring should provide representative samples for statistical evaluations. The public can play an important role in evaluating damage to visible elements of the architectural heritage and whether this constitutes damage to envelope structures or decorative elements, and to review the benefits and costs of any intervention.

In Fig. 8.22 we can follow an intervention history of a sculptural monument where the periods between different actions are shown – both maintenance and more substantial interventions – which is another significant parameter for the establishment of a heritage management strategy when combined with financial or other benefits.

When dealing with costs, it is useful to distinguish between different types of actions and itemise them accordingly. It seems practical to consider three modes of intervention, even though we can define more subtle divisions. The three are: maintenance, conservation, and renovation. The selected categories correspond to typical demands for individual operations and their combinations which are reflected in the relevant costs (and lifetimes). All three groups can be further divided according to specific challenges and demands coming from the need to safeguard and protect various grades or levels of cultural heritage values which may require specific skills, technologies and materials.

According to the ICOMOS United Kingdom Charter for the conservation of buildings and sites, maintenance is defined as "the routine work necessary to keep the fabric of a building, moving parts of machinery, grounds, gardens or

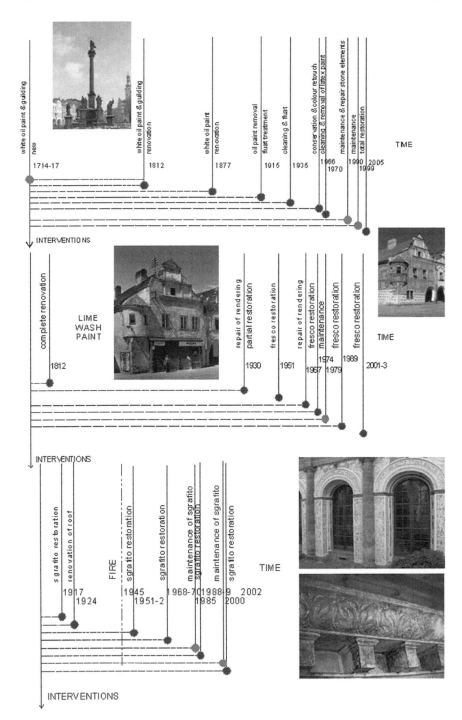

Fig. 8.22 Return periods of typical restoration and maintenance works

any other artefact in good order". Among the maintenance actions we consider mainly

- regular inspection
- cleaning including biocide treatment
- local repair of the fabric of a building or artefact
- renewal of protective layers (water repellent treatment, paints, ...).

Conservation is then defined as "action to secure the survival or preservation for the future of buildings, cultural artefacts, natural resources, energy or any other thing of acknowledged value. It is considered as the action taken to prevent decay and manage change dynamically". It is a quite complex definition, but usually it can be restricted to the following actions:

- survey and research
- maintenance
- consolidation
- restoration.

The last of the modes of intervention are grouped together under the label "renovation" and mainly include

- substantial repair
- alteration
- rebuilding or remaking
- replication.

Under the term repair we include work to the fabric of a building or artefact to remedy defects, significant decay or damage caused deliberately or by accident, or by neglect, normal weathering or wear and tear. The objective is to return the building or an artefact to good order, without too great an amount of alteration and restoration.

The next figure (Fig. 8.23) shows an illustrative comparison of the typical modes of interventions mentioned above in terms of costs and lifetimes. The CULT-STRAT project reviewed costs of construction works or building elements related to appropriate units – square metres, linear metres, and typical elements. The prices include the material costs and labour costs with a mean transportation and scaffolding costs relevant for low rise buildings up to 3 storeys. The costs are related to the year 2005 and are derived from known examples (mostly restoration costs), official price lists (special & general construction works) and long-term statistical surveys (general repair & construction costs).

Economic data on conservation and maintenance, namely the costs of conservation and maintenance works, provide a basis for optimum planning of interventions on cultural heritage objects. The typical average costs of restoration and maintenance works in European countries differ substantially, even when corrected to local costs of living (see below). The scope of costs associated with roof maintenance, conservation and renovation works are clearly seen in Fig. 8.24 where costs of ten selected actions

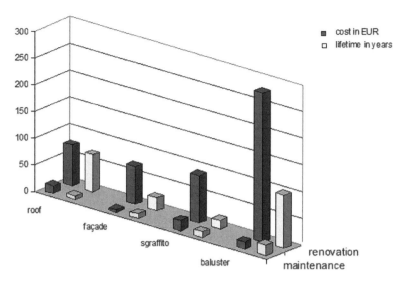

Fig. 8.23 Comparison of costs and life times for selected renovation and maintenance

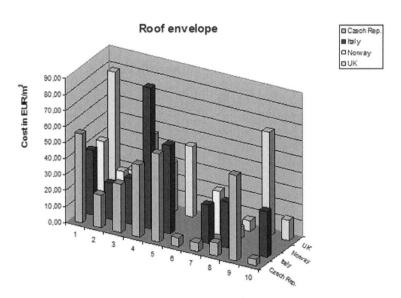

Fig. 8.24 Costs reduced by the ratio corresponding to the PPP in the relevant country

are presented – in this specific case they are reduced by the ratio corresponding to the purchasing power parity (which relates the relative values of different economies by comparison of what they are able to purchase) in the relevant country and, therefore, the influence of differences in local economies are removed. There are still apparent significant regional differences which are mostly caused by the cost of labour in the respective countries. This is an important fact as it may also influence approaches to the establishment of heritage management strategies.

Figure 8.23 refers to works of the following specification:

- Roof wooden shingle (painting for maintenance data)
- Façade lime mortar simple plaster (baroque style)
- Sgraffito (hydrophobization for maintenance)
- Baluster sandstone (cleaning for maintenance)

The costs, combined with the lifetime assessment mentioned above, yield a formula for computing a general cost for pollution. The life assessment can be calculated from the damage functions or known from practice. Of course, the result is further dependent on the situation in individual countries. First, variations in the types of materials used for architectural heritage objects across Europe are relatively small, but variations are quite high in the volumes of materials typical for different regions. The volumes of various historic materials under threat are further influenced by the policies in different countries for listing buildings and structures as architectural heritage and the degree of protection that the listing gives.

Although questions of cost benefit have only been partially answered so far, an intelligent evaluation, of pollution monitoring data, data on maintenance costs and life, to necessary maintenance estimates, supports the establishment of reasonable heritage maintenance policies at municipality level. Such an approach seems to be realistic and viable, thanks to better availability of data both on materials exposed to pollution and on local pollution sources in cities and villages. The results help to adopt decisions on maintenance or restoration planning, and to evaluate the effectiveness of interventions that have been made.

Figure 8.24 refers to the following works

1. copper sheet replacement
2. galvanized steel replacement
3. plain tiles (double) replacement
4. flap pantile replacement
5. wooden shingle split replacement
6. paint on steel sheet renewal
7. paint on galvanized steel sheet retention
8. paint on wooden shingle
9. slate tiles
10. flap pantile 10% repair

8.7 Cost/Benefit Analyses for Selected Built Heritage Types

The cost/benefit analysis of environmental strategies related to individual cultural heritage objects represents a difficult task and it is distorted by so many side effects, constraints and conditions that only very approximate or general tendencies can be revealed and used as the basis for drawing conclusions.

This book has focussed on environmental damage but making decisions on both the conservation regime of monuments and building maintenance is influenced by many other factors, including

- the availability of financial resources;
- the importance of the monument from different points of view (e.g. architectural, historical, artistic, etc.);
- the quality and expert knowledge of technical supervision of a monument or building (this can include knowledge of the causes of the specific technical problems and up-to-date repair and protective technologies which can be straightforward but often require special materials and technologies, not that common in building industry);
- the extent of previous building or monument maintenance, which may not have been done regularly and with the same quality during the life of a historical object;
- the materials used and complexity of the object's surface – leading to a greater demand for maintenance, and material diversity resulting in different ageing as well as different durability or resistance against weathering.

The ratio of sheltered to openly exposed parts of the monument or building influences staining and leads to uneven soiling, which is a problem particularly for monuments bearing carving and relief with a high artistic value or other surface value, the loss of which would be considered priceless. In general, the appearance of architectural heritage and sculptures represent the main phenomenon which can be affected by controlling air pollution and maintenance strategies and policies. In cases where the surface is considered priceless it is not possible to tolerate any material loss and the life time estimates will concern only protective measures, mostly protective layers or treatment. In other words, the measures protecting highly sensitive and precious cultural heritage from surface degradation and material loss are to be applied, maintained and renovated in any damaging environment regardless of its corrosivity. For the cost/benefit analysis in such cases the damage functions and lifetime estimates of protective measures are indispensable.

8.7.1 Case Studies of Typical Elements of Built Heritage From the Cost/Benefit Point of View Taking Into Account Pollution Situations

The CULT-STRAT studies have been performed to illustrate the application of many of the techniques and measures discussed in this volume to local situations. They used a standard methodology including the following items:

A. Basic data

Position, importance, historic review, architectural styles, artistic elements, decoration, function(s), use ...

B. Material & Structural data ("modulus" description)

Illustrated (documented) description of materials used, structural systems, protective elements and systems, area of individual materials (in %), exposition to external loads and actions ...

C. Environmental data

Environmental situation, air pollution, material dampness, winds, temperature, moisture (RH), particles, development and tendencies.

D. Structural health condition

Illustrated (documented) description of contemporary state, deterioration, decay and damages and their short analyses ...

E. History of interventions

Historical review of maintenance, restoration, reconstruction, adaptation, conservation and similar interventions – with description of the extent and intensity, lifetimes of individual materials, elements and systems, with graphical representation of tendencies ...

F. Costs

Historic review of direct costs related to the above interventions, broken down (if available) according to individual materials, elements, technologies and pollution degradation etc. & indirect costs (limited here only to value to visitors and general public) ...

G. Benefits

Review of benefits to the owner from the use, e.g. tourism, including possible indirect benefits from employment, local trade etc. (only related to the analyzed site)

A detailed search for environmental, economy and life cycle data has been performed in three different European towns – Tel, Prague and Paris. Examples of the results are presented in Figs. 8.25–8.27. This type of inventory studies can serve as an important support for decisions about planning of maintenance periods for the materials and objects discussed. The case studies suffer from the fact that complex and mutually correctly linked environmental, economy and lifetime data are often not easily

IV.1.2.8   Prague: Metallic (bronze) sculpture – St. George fountain(Prague Castle)

Basic data:
The sculpture is a replica of the original
statue. The replica was created in 1966,
the original statue in 1373. The granite
plinth was built in 1929.
Environmental data:
The object is located in urban environment
(large town), accessible to tourists. The
whole surface is soiled with dust, organic
particles and corrosion products.
Health conditions:
Relatively good condition, regularly
maintained. Metal parts corroded.

History of intervention:
The sculpture with bronze basin was
restored in 1995. Bronze was washed with
pressure water with degreasing agent, the
released patina (oxides and sulphides) was
softly brushed away and the surface
conserved by wax.
In 1996 maintenance treatment of the
sculpture only with washing the bronze
with pressure water with degreasing agent,
the released patina was softly brushed
away and the surface conserved by wax.

Affected surface area: (estimated value)
                                          31,25 m^2
Cost of the intervention: (2006)
                                   22 967,00 CZK
Restoration in 1995 of a larger area cost
                                    42 000,00 CZK
Cost per unit
                                      735 CZK/m^2
Life cycle of a substantial restoration:
                                               years
Life cycle of maintenance: (real)
                                           11 years

Cost/benefit comments:
The influence of environmental impact
negligible, tinned iron slightly corroded.

Fig. 8.25 Environmental, economy and life cycle data for metallic bronze sculpture,
St George Fountain, Prague Castle

available and a part of them would need to be collected from monitoring
stations. The acquired information therefore reveals some features of return
periods but it is difficult to relate them to the environmental – air pollution
situation.

IV.1.1.2   Gothic building–st. Jacob's church lime mortar façade

Basic data:
Original church founded and built between 1360–1370. It burned in 1368 and repaired after the year 1443. The tower covered with the present roof in 1687.

Materials and structural data:
Mixed masonry is plastered and stone architectural elements exposed. Lime mortars 1271 sqm (new plaster 879 sqm and 392 sqm repaired), flat ceramic roof tiles (996 sqm), granite (343 sqm), copper sheets (131 sqm), iron parts (14 sqm), artifical stone (11 sqm), wood (2 sqm).

Environmental and climate data:
The building is located in a clean urban environment (small town). Altitude 521 m, in winter possible inversion situations up to 20 m above the ground, average year temperature 6.5°C, average precipitation 617 mm, average number of days with snow cover 71,7, average number of days with temperature dropping below 0°C 141,3, average number of summer days with maximum day temperature over 25°C 47,4.

Structural health condition:
Detachment and spalling of the rendering (up to the height of about 1 m). Completely missing on window parapets. The moisture of plaster is low and the salinity negligible. Structural problems of masonry tower fixed in sixties.

History of interventions:
In the recent history the church was restored substantially in 1891 (purification of the gothic shape) – restoration of presbytery, alteraation of windows, new rendering, relaying of roof tiles, roof framing repair, 1924 façade rendering repair and renovattion of paints, cleaning of quoin stones, 1940 repair of rendering on presbytery, 1948 repair of rendering, cleaning of buttreses, quoin stones and portal, 1949 the first stage of renovation terminated, 1960–61 experts opinions declare dangerous deformation of masonry tower with apparent cracks (due to a stone quarry blasting), 1962–67 strengthening and reinforcement of the church tower, 1975–76 a complete renewal of interior as well as exterior plastering, 1984 repair of rendering of tower, 1995 roof tile replacement (presbytery and church), 1997 new copper sheet covering of the bottom part of the tower roof, 2001–2 total renovation of façade (lime mortar and copper flashing).

Affected surface area: (estimated value)

Fig. 8.26 Environmental, economy and life cycle data for Gothic Building, St Jacob's Church, Telc

IV.1.4 Paris. Place de la Concorde (n°15-16-17-18 on Paris Map)

Basic data: The largest square in Paris. When royal architect Jacques-Ange Gabriel designed the Place de la Concorde, he made sure that bulidings would face only its North side and, in 1757, work began on the two matching Neoclassical North façades separated down the middle by Rue Royale (1). The works were completed in 1770. The Hôtel de la Marine (now Ministry of the Navy) (2)(3) is situated on the Eastern side and the Western one comprises the Hôtel de Coislin (4), the French Automobile Club (5) and the very elegant Hôtel de Crillon (6).

Material and structural data:
Lutetian Parisian limestone from Paris surroundings (Coflans Sainte Honorine).

Environmental data:
One of the busiest square in Paris. High level of air pollution.

Structural health condition:
Good.

History of interventions:
Last interventions by soft dry sand-blasting in 1985.

Affected surface area:
Ministry of the Navy: 8 000 m^2
Hôtel de Coislin: 2 880 m^2
French Automobile Club: 1 440 m^2
Hôtel de Crillon: 2 920 m^2

Cost of the intervention:
Ministry of the Navy: 224 000 €
Hôtel de Coislin: 80 640 €
French Automobile Club: 40 320 €
Hôtel de Crillon: 81 760 €

Cost per unit: €/m^2

Life cycle of a substantial restoration: 40 years

Life cycle of a maintenance:?

Fig. 8.27 Environmental, economy and life cycle data for Place de la Concorde, Paris

8.8 Damage Caused by Air Pollution to Indoor Cultural Heritage Materials

Another area of recent research that is extremely relevant to local scale management seeks to evaluate the indoor situation.

Czop (2002) divides the sources of indoor pollutants into four main groups:

- External pollution, mainly from industry and traffic (air pollutants and airborne particles).
- Internal pollution from heating systems, man made pollution (dust).
- Contamination from/of materials – wood (oak), acidic paper, PCV, fibre-board, PVA (acetic acid).
- Accident, vandalism.

The first two groups in this list are likely to contain gases and particles consistent with those normally considered damaging to the external surfaces of cultural heritage – sulphur dioxide, oxides of nitrogen, ozone, nitric acid, and particulates. These are all likely to interact with materials (metals, glass, stone, etc.) by mechanisms similar to those experienced externally. But there is an expectation that the rates of decay will be much lower because of the lower pollutant concentrations, limited amount of water and low relative humidity. This is likely to be true in museums where the environment can be controlled but very much less likely to be true in buildings such as churches or where the building is a ruin. However, there is published evidence that deterioration does occur – both corrosion of metals and surface loss from stone in museum environments.

In addition, Czop (2002) mentions the contribution from post-conservation contamination (for example, solvent vapours, products from plastic materials: plasticisers, products of depolymerisation, improper materials), maintenance pollution (dusting, floor and furniture polishes and cleaners, wall and furniture paints, window cleaners (ammonia, organic acids), visitors (VOCs (volatile organic compounds) from cosmetics, dust). However, these all seem more likely to relate to the care and maintenance of historic objects and artefacts than to the materials used in the building interior.

Indoor air pollution has also been associated with deterioration of cultural heritage objects (Tetreault, 2003; Blades et al., 2000; Brimblecombe, 1990). Degradation of materials is of great importance in the case of museums, historic buildings and archives. Data on historic concentrations of sulphur dioxide measured inside the British Museum, London during the period 1850–2000 shows that the highest concentrations were measured between 1880 and 1930 when they were over 20 $\mu g\ m^{-3}$, but this was still only around 10% of the external concentrations. The concentration then declined rapidly until by 1980 they were less than 2 $\mu g\ m^{-3}$ (Brimblecombe, 2006). Brimblecombe (2006) also presented data on concentrations of nitrogen dioxide found at different locations within a museum and compared them with external concentrations.

A reduction of around 50% between external and internal was found, with a still greater reduction further into the naturally ventilated building which is attributed to deposition of the gases by the internal fabric of the building. Deposition can be seen as a cumulative process but it is likely to remain a surface process as there is limited moisture to transport pollutants or reaction products further into the materials. However, there does seem to be sufficient pollutants and moisture from variations in relative humidity to result in corrosion of metals – though the more extreme examples of corrosion seem to be linked to chemicals from the materials around specific objects, for example acids from wood or paper.

8.8.1 Modelling of Indoor Air Quality and Pollutants Infiltrated from Outdoors in Museums and Galleries.

Extensive research effort has been invested in examining the factors that determine the concentration of pollutants indoors and the indoor air quality. These are primarily the introduction of ambient air through the infiltration of outdoor air indoors, the emission of pollutants directly to the indoor air by indoor sources and their removal by deposition on indoor surfaces, and the heterogeneous (on indoor surfaces) and homogeneous (gaseous phase) chemical reactions (Weschler and Shields, 1997; Ekberg, 1994).

In the absence of significant indoor sources the concentration of pollutants in indoor air varies proportionally with the concentrations in outdoor air, and the indoor air can be considered as an extension of the outdoor (Jones, 1999). The influence of the outdoor air quality on the indoor air quality is dependent on the climate and the building design. Meteorological conditions play an important role by determining the concentration of pollutants outdoors and also the natural ventilation rate (wind speed, temperature and pressure gradients). The building design and construction materials affect the transport of pollutants between different chambers inside the structure, the deposition velocity of indoor surfaces and the infiltration of the outdoor air indoors through openings in the building shell (Lazaridis et al., 2006). The room geometry, described in a simplified form by the A/V (area/volume)-ratio, determines the deposition rate relative to the ventilation rate. A high infiltration rate and indoor emissions give higher indoor concentrations, a high deposition velocity to indoor surfaces and a higher A/V-ratio give lower indoor concentrations. The effects of hetero- and homogeneous reactions have to be evaluated separately for each particular reaction and component.

Many different models have been developed to examine the influences of the range of factors that determine the indoor air quality, and to estimate indoor air pollutant concentration and exposure. The different approaches include mass balance, empirical – semi-empirical models and models based on

Computational Fluid Dynamics (CFD). More specifically, dynamic models are based on mass balance equations for describing the fate of pollutants in the indoor air (Dimitropoulou et al., 2001; Nazaroff and Cass, 1986; Hayes, 1989). These models account for the infiltration of outdoor air indoors, the emission by indoor sources and production/removal by chemical reactions. In addition their application requires experimentally resolved values on the air exchange rate and the room-mixing factor in order to adequately estimate the concentration of pollutants indoors (Chaloulakou and Mavroidis, 2002). The deposition velocities or kinetic coefficients used are usually mean values obtained from literature, but can also be experimental estimations of deposition velocities for different kinds of materials (IMPACT, 2006; Grøntoft and Raychauduhri, 2004; Liu and Nazaroff, 2001; Cano-Ruiz et al., 1993; Brimblecombe, 1988). The rooms are considered to be rectangular well-mixed boxes and to some of these models the exchange of air between indoor microenvironments is considered (multi-chamber models, Nazaroff and Cass, 1986; Hayes, 1989; Dimitropoulou et al., 2001). These models can only be applied for well-mixed environments where the concentration of pollutants is homogeneous throughout the room.

Semi-empirical models are based on large data sets from field measurements. Their application does not require air exchange rate measurements, but their use is limited to a specific experimental interval of environmental conditions and pollutant concentration values (Milind and Patil, 2002; Thatcher and Layton, 1994).

CFD models solve equations derived from mass conservation conditions to estimate the spatial distribution of pollutants concentration indoors (Chen et al., 2006; Fan, 1995; Hayes, 1991). Their main disadvantages are that deposition rates used are usually empirically estimated or in other cases ignored (Chen et al., 2006) and they are incapable to handle mixed natural-forced airflow and simulate the occupant-behaviour related factors (Fan, 1995).

Outdoor/Indoor models have been applied to museums particularly for the estimation of indoor O_3 concentration (Salmon et al., 2000; Papakonstantinou et al., 2000; Druzik et al., 1990; Nazaroff and Cass, 1986). More specifically Salmon et al. (2000) and Druzik et al. (1990) applied the mass balance model of Nazaroff and Cass (1986) to estimate the O_3 indoor concentration in several museums in the historic central district of Krakow, Poland and 11 museums in the areas of Los Angeles and San Diego California, USA, respectively. The model of Nazaroff and Cass (1986) has been validated with experimental data in different indoor environments including museums. The model has also been used to evaluate the impact of different preventive strategies for the protection of museum collections from damage to atmospheric ozone (Cass et al., 1991). Papakonstantinou et al. (1999) developed a CFD model and applied it in the archaeological museum of Athens. However the model has not been validated with experimental data. Measurements conducted in several museums (Gysels et al., 2004; Brimblecombe et al., 1999; Camuffo et al., 2001) demonstrated that the concentration of pollutants do not vary significantly within a room and between adjacent interconnected rooms. Mass balance models can be efficiently applied in the case of museums, historic buildings and archives (Lazaridis et al., 2006).

8.8.2 Exemplifying Models – The IMPACT Model. A Simple One-Zone Mass balance Model for Use in Museums and Archives

The IMPACT model, from the IMPACT (Innovative Modelling of Museum Pollution and Conservation Thresholds) Project is a web-based software tool designed to predict indoor concentrations of the most damaging gaseous pollutants, that origins from outdoors, found inside museums and historical archives (Fig. 8.28). The model is a Java Applet accessible via the Internet and can be used without special license. The IMPACT model was developed as part of the EU – IMPACT Project (EVK4-CT-2000-00031) and tested with field test data (Grøntoft et al., 2005) from the EU project "MASTER" (MASTER, 2006).

The IMPACT model can be used to calculate the indoor average concentration of NO_2, O_3 and SO_2. The behaviour of these gaseous pollutants is governed by a mass balance equation of the form:

$$\frac{dC}{dt} = P\lambda C_o(t) - \lambda C_i(t) - v_d \frac{A}{V} C_i(t) - kC_i(t) + S(t)$$

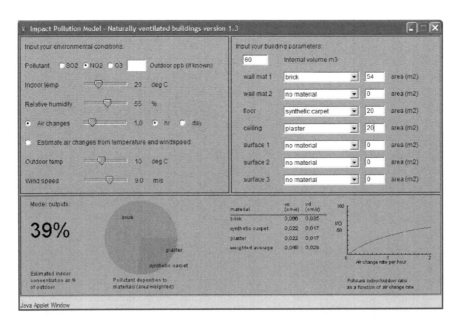

Fig. 8.28 The web interface for the IMPACT project – naturally ventilated buildings model

In the above equation, the first term on the right hand side represents the portion of the outdoor concentration that enters the indoor environment, the second, third and fourth terms represent pollutant losses due to exfiltration, deposition and chemical reactions respectively and the last term represents the pollutant production from indoor sources.

The model is an equilibrium model, that is, it assumes no change with time, $(dC/dt = 0)$. The model assumes that there are no indoor sources of the gases it models $(S = 0)$, NO_2, O_3 and SO_2, and that pollutants are chemically inert $(k = 0)$, and that the filteration factor $p = 1$. Moreover measured values for net deposition to indoor surfaces are used, thus avoiding the problem of desorption of pollutants from the materials. The model takes into account the influence of temperature and relative humidity on deposition velocity. Thus the IMPACT model calculates the indoor average concentration of NO_2, O_3 and SO_2 by solving a deposition-based mass balance equation:

$$\frac{C_i}{C_0} = \frac{\lambda}{\lambda + v_d(A/V)}$$

where:

C$_i$ and C$_0$ are the pollutant concentration indoors and outdoors,
λ is the overall building ventilation rate (air exchange rate, hr-1),
A/V is the surface area to volume ratio of interior $((m^2)/(m^3))$ and
v_d is the deposition velocity (m hr^{-1} or cm s^{-1}), an expression of how well a
    particular surface takes up a particular pollutant gas.

In the IMPACT model there is no dependence of concentration on time and mean concentrations for long periods of time can be calculated with the model. In most cases, long-time average concentrations are more important when it is desirable to estimate deposition of pollutants to materials related to the deterioration of the art works. Short-term elevated concentrations might cause problem to humans and be a threat to human health, but do not contribute much in the deterioration of art works exhibited in museums, historic buildings and archives.

Since indoor emissions are considered to be zero in the model, pollutants can be transported indoor only by infiltration from the outdoor environment through open doors and windows and through cracks of the building shell. The study area is modelled as a rectangular box, which communicates with the outdoor environment via airflow. The whole room is treated as a single well-mixed zone and the concentrations of the gaseous pollutants are assumed to be uniform throughout the room. Pollutants are removed by exfiltration and deposition on indoor surfaces. Deposition values for different materials used in the model have been estimated from intensive laboratory measurements as part of the IMPACT project. The user can choose the material covering indoor surfaces such as walls, ceiling, floor and other large objects (e.g. showcases)

Table 8.1 List of materials for indoor surfaces, used in the IMPACT model

Brick	Glass	Plastic	Wood, oiled
Cardboard	Granite	Sandstone, calcareous	Wood, hard
Chipboard	Limestone	Sandstone, silicate	Wood, painted
Cloth	Marble	Slate	Wood, soft
Carbon cloth	Metal	Synthetic carpet	Wool textile
Concrete, coarse	Paintings	Synthetic floor	
Concrete, fine	Plaster	Wallpaper	

found inside the room. Table 8.1 presents the list of materials that are available in the room.

The air exchange rate can be entered directly in air exchanges per hour or air exchanges per day, the latter being more suitable for museum display cases, or roughly estimated by the difference in temperature between the inside and outside of the building and the external wind speed. The model can be applied both in naturally ventilated and mechanically ventilated buildings. In a mechanically ventilated building, the air entering the room is a mixture of fresh air from outside and recirculated indoor air that has been purified by a combination of mechanical filters. The user is asked to give values regarding the air intake, the filter efficiency and the portion of fresh air to recirculated air entering the indoor environment. In equilibrium conditions the mass balance equation used is (EU project IMPACT, 2004):

$$\frac{C_i}{C_o} = \left(\frac{(1 - \eta)f_{ox} + f_{oi}}{f_{io} + v_d A + f_{ix}\eta} \right)$$

where

f_{ox} is the fresh air intake to the mechanical ventilation system,
f_{ix} is the quantity of air, which is recirculated,
f_{io} is the exfiltration/mechanical exhaust from the building,
f_{oi} is the natural infiltration,
η is the filter efficiency,
A is the surface area of the room.

A schematic representation of infiltration conditions in the case of mechanically ventilated buildings is displayed in Fig. 8.29, a schematic representation of infiltration conditions in the case of mechanically ventilated buildings. In the limit of the mechanical airflows being zero, the model gives exactly the same answers as the equation used for naturally ventilated buildings.

A semi-empirical algorithm was developed in order to model the influence of temperature and relative humidity on deposition velocity. The algorithm was applied in combination with laboratory measurements in selected values of temperature and relative humidity. The deposition velocity for the temperature and humidity ranges 0–35°C and 0–100% accordingly was found by

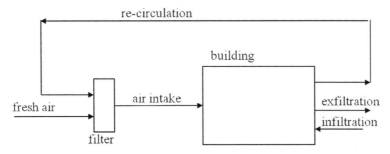

Fig. 8.29 A schematic representation of infiltration conditions in the case of mechanically ventilated buildings

interpolation between the selected values. The user of the model can introduce the mean temperature and relative humidity for the selected modelled period.

8.8.2.1 An Evaluation of Results Using the IMPACT Model

An evaluation of the IMPACT model was performed with extensive field data obtained in a one-year measurement campaign during the EU – project MASTER (MASTER, 2006). In this campaign the parameter values relevant for the modelling; monthly mean values for outdoor and indoor concentrations of the gases SO_2, NO_2 and O_3, temperature and relative humidity were collected. Ventilation rates, which is usually the less available of the needed parameters was simulated in the modelling using measured outdoor to indoor concentrations of the least reactive gas, NO_2. The evaluation showed that NO_2 could be quite well modelled using the IMPACT model, but that the model tended to overestimate indoor O_3. Model evaluation with SO_2 was not possible due to low outdoor and very low indoors concentrations almost equal to zero in all the museums.

It was concluded that a minor reason for the overestimation of O_3 could be underestimation of the real indoors areas used as input to the model. The main reasons for the overestimation of O_3 was that much of the O_3 was removed before it entered the room where measurements were performed, and that this was not properly considered by the modelling procedure and model specifications. The modelled room usually had ventilation from adjacent rooms and the "outdoor" concentration that should have been used in the modelling was partly that of these adjacent rooms rather than the higher outdoor concentrations, which were measured and used in the modelling. Also it is known that ozone shows very high reaction probability (γ) for cracks, with crack height less than 0.5 mm. Ozone deposits strongly in the surfaces of the cracks and its penetration to the indoor environment is reduced. (Nazaroff and Liu, 2001). This is not accounted for in the IMPACT model which should be improved with the inclusion of a penetration coefficient for ozone. It would however be very difficult for the user to assess this parameter. Homogeneous chemistry may also influence indoor O_3 concentrations. Ozone is known to react in the relative darkness indoors with NO that infiltrates from traffic emissions, especially during the summer. It also reacts with Volatile Organic Compounds (VOCs), especially terpenes that may be present indoors. Indoors, O_3 is easily overestimated in relatively simple models due to the lack of inclusion of all these important sinks.

Pollution modelling for museum and archive indoors can thus be used to evaluate pollution dose exposure of interiors and objects. It can further be used for evaluating different ventilation scenarios and ventilation system designs and for estimating the air exchange rate that prevents indoor concentration of O_3 and other oxidizing pollutants to exceed acceptable concentrations for preservation. Moreover such models can provide decision makers with valuable information considering the emission abatement strategies in the areas in the vicinity of the cultural heritages sites, museums and historical archives. It is however important to have some knowledge about specific pollutant characteristics and model specifications to be able to consider possible model bias.

8.9 Strategies to Mitigate Outdoor Pollution Effects on Indoor Environments

This short section extends the discussions in the previous sections to the impact of pollutants that originate in the external environment but which present a risk to museum collections and to the interior fixtures and fittings of the built heritage. This is a major subject and has been included in a number of EU funded projects, for example IMPACT (Innovative Modelling of Museum Pollution and Conservation Thresholds) which dealt with the development of tools in order to contribute to a better preventive conservation strategy. There are also many specialist publications on the subject of museum environments and their effect on artefacts and fittings, and amongst these are the publications from the IAQ (Indoor Air Quality) conferences.

8.9.1 Control of Air Pollutants in the Indoor Environment

Avoiding polluting sources is the key issue here. It is always better to avoid air pollution in the first place than to be forced to make all sorts of mitigating arrangements later – with the possibility of losing objects in the meantime due to deterioration, or having to take on costly conservation treatments.

There are two strategies for controlling pollutants in internal environments – active and passive strategies (Czop, 2002).
Active:

- Ventilation control – introduction of dilution air to reduce contaminant levels below threshold levels. This is most likely to be successful where source of the pollutant is internal and where the "outside" air is not too polluted.
- Removal control – air filtration media: absorption – activated carbon, chemisorption – potassium permanganate on activated alumina (Al_2O_3), filters for airborne particulates.

Passive:

- Source control – optimal, but not easy, since sources are difficult to locate and particularly difficult if the sources are external and distant.
- Removal control – pollutant absorbents (ZnO to prevent silver from tarnishing).

Active control would appear to be the more effective method but it may not be at all practical in heritage buildings and structures where air movement is, of necessity, uncontrolled.

8.9.2 "Tolerable" Levels for Air Pollutants and Material Damage in Indoor Microclimates

The concept of "tolerable" concentrations of pollutant can be applied to indoor environments in the same way that it is applied to outdoor environments (which will be discussed in the next chapter) and the arguments as to what is tolerable appear to be very similar. In cases where objects are considered extremely important or are susceptible to decay then the tolerable deterioration rate may be considered to be zero and by inference the tolerable concentration of the one or more of the key parameters must also be zero. For example, it may be necessary to reduce sulphur dioxide to zero if the humidity cannot be controlled such that the gas cannot react.

The need for the "complete" protection of some artefacts has resulted in the development of different corrosivity systems and classes. In the ANSI/ISA system for environmental conditions for process measurements and control systems the corrosivity classes range from S1 where the air is required to be "extremely pure" to S4 where there can be "slight contamination" (ANSI/ISA 1985). Class S1 is suggested for archives, metal collections and rare books (Muller et al., 2006). The classes are based on the corrosion of copper and silver exposed as "corrosion coupons" with each class having a maximum surface corrosion rate over a 30-day period.

Within the ISO TC 156 a more extensive system has been developed for classification of low corrosivity in indoor atmospheres (ISO 11 844). According to Part 1 of the standard the corrosivity is classified in five categories denoted from IC 1 (very low corrosivity) to IC 5 (very high corrosivity). The determination of corrosivity is based on measurements of corrosion attack on standard coupons of four reference metals – carbon steel, zinc, copper, and silver – after an exposure of one year. From the mass loss or mass increase the indoor corrosivity category towards each metal is determined. The standard also gives a relation between the ISO and ISA systems. The ISO standard also contains guidelines for estimation of indoor corrosivity based on the knowledge of humidity, temperature and pollution characteristics of the given environment and a table describing typical environments related to the indoor corrosivity categories. Part 2 describes the methods for determination of corrosion attack on metal specimens and Part 3 the measurements of environmental parameters affecting indoor corrosivity.

8.10 Conservation & Maintenance Strategies

8.10.1 General Principles

Buildings and monuments should be considered within the context of the overall built heritage and so repair and restoration should be seen as a conservation project within an overall conservation plan which draws on an underlying conservation philosophy.

The key considerations for a conservation philosophy can be summarised as:

- An holistic approach – the buildings and their setting needs to be considered as an entity;
- The historic environment must allow for new architecture and modern lifestyles – but with a presumption in favour of preservation;
- Conservation and restoration must have recourse to all the sciences and techniques which can contribute to the study and safeguarding of the architectural heritage;
- Restoration and refurbishment should be based on a programme of "minimum intervention" whilst taking into account the need to balance the costs of such work and the disruption of activities on the site;
- Wherever possible interventions should be reversible – that is that they can reversed at a later date, for example if a better solution becomes available at some future date.

In the case of the many aspects of the built cultural heritage this means looking for remedial techniques and materials that will prolong the service life of the stone (and the building) whilst considering the implications of current work for future maintenance and interventions. For example, the considering of the consequences of the application of a water repellent treatment on future cleaning or the application of any other water-based treatments.

It is possible to identify a hierarchy of guidance relating to conservation and maintenance strategies and these in turn inform the practitioner of the techniques which are applicable in different circumstances. At the top of the hierarchy are conservation principles and these are summarised in a number of "charters" (for example the Burra Charter and the Stirling Charter) that can be summarised as:

- The historic environment is a shared resource
- It is essential to understand and sustain what is valuable in the historic environment
- Everyone can make a contribution
- Understanding the values of places is vital
- Places should be managed to sustain their significance
- Decisions about change must be reasonable and transparent
- It is essential to document and learn from decisions

(Taken from Conservation Principles for the Sustainable Management of the Historic Environment, English Heritage, 2005)
   In this case the following definitions have been used:

Conservation:

The process of managing change in ways that will best sustain the values of a
   place in its contexts, and which recognises opportunities to reveal and
   reinforce those values
Historic environment:

   All aspects of the environment resulting from the interaction between people and places through time
Significance:

The sum of the cultural and natural heritage values of a place
Value:

   An aspect of worth or importance, in this context, ascribed by people to places, where value can be further broken down into:

- Aesthetic Value: Relating to the ways in which people respond to a place through sensory and intellectual experience of it
- Community Value: Relating to the meanings of a place for the people who identify with it, and whose collective memory it holds
- Evidential Value: Relating to the potential of a place to yield primary information about past human activity
- Historical Value: Relating to the ways in which a place can provide direct links to past people, events and aspects of life
- Instrumental Value: Economic, educational, recreational and other benefits which exist as a consequence of the cultural or natural heritage values of a place.

(Adapted from Conservation Principles for the Sustainable Management of the Historic Environment, English Heritage, 2005)
   One of the most relevant documents to refer to when considering the conservation and maintenance of the built environment typical of many urban

landscapes is the International Committee on Monuments and Sites (ICOMOS) Charter on the Built Vernacular Heritage (1999). The introduction states:

> The built vernacular heritage occupies a central place in the affection and pride of all peoples. It has been accepted as a characteristic and attractive product of society. It appears informal, but nevertheless orderly. It is utilitarian and at the same time possesses interest and beauty. It is a focus of contemporary life and at the same time a record of the history of society. Although it is the work of man it is also the creation of time. It would be unworthy of the heritage of man if care were not taken to conserve these traditional harmonies which constitute the core of man's own existence.
>
> The built vernacular heritage is important; it is the fundamental expression of the culture of a community, of its relationship with its territory and, at the same time, the expression of the world's cultural diversity.
>
> Vernacular building is the traditional and natural way by which communities house themselves. It is a continuing process including necessary changes and continuous adaptation as a response to social and environmental constraints. The survival of this tradition is threatened world-wide by the forces of economic, cultural and architectural homogenisation. How these forces can be met is a fundamental problem that must be addressed by communities and also by governments, planners, architects, conservationists and by a multidisciplinary group of specialists.

8.11 National and Regional Organizations and Guidance

Examples of national and regional guidance and the legislative organisations established to undertake are given here to illustrate the relationship between local maintenance strategies and prevailing heritage policy.

8.11.1 Legislative Organisations – UK

More detailed guidance on the conservation legislation and techniques can be found at national and regional level. For example in the UK, the following Government agencies & Royal Commissions have a special role in maintaining cultural heritage:

- English Heritage (the Historic Buildings and Monuments Commission for England) is the national body created by Parliament in 1984 charged with the protection of the historic environment and with promoting public understanding and enjoyment of it. It is the Government's official adviser on all matters concerning heritage conservation, and provides substantial funding for archaeology, conservation areas, and the repair of historic buildings, churches and cathedrals. It is also responsible for some 400 historic properties in the nation's care. The web site includes details of the Archaeology Division and their Commissions, and links to a number of online publications, such as the Archaeology Review, the Management of

Archaeological Projects document, their Research Agenda, Monument Class Descriptions, and their Ancient Monument Laboratory's geophysical survey database. In April 1999, English Heritage was merged with the Royal Commission on the Historical Monuments of England, and the new merged organisation incorporates the English National Monuments Record (NMR).

- Historic Scotland – An Executive Agency of the Scottish Office that safeguards and promotes understanding and enjoyment of ancient monuments, archaeological sites and landscapes, historic buildings, parks, gardens and designed landscapes. Schedules monuments of national importance; cares for and opens over 300 monuments to the public. Monitors applications to alter or demolish listed buildings, and with advice from the Historic Buildings Council for Scotland gives grants for their repair.
- CADW: Welsh Historic Monuments – Created in 1984, CADW carries out the complete range of responsibilities for the conservation, presentation, and promotion of the built heritage of Wales on behalf of the National Assembly for Wales.
- Environment & Heritage Service, Northern Ireland – The Environment and Heritage Agency within the Department of the Environment for Northern Ireland with responsibility for the protection, recording and conservation of monuments and buildings in Northern Ireland and aims to enhance public awareness by publicity, publications and education. It is also responsible for scheduling historic monuments and listing buildings of special architectural or historic interest. There are some 180 monuments in state care. Other duties include excavation, recording the built heritage, the NI Monuments and Buildings Record, publication and education.
- Royal Commission on the Ancient & Historical Monuments of Scotland – The Commission is an independent non-departmental government body financed by Parliament through the Scottish Office under the sponsorship of Historic Scotland. The main objectives of the Commission are: to record and interpret the sites, monuments and buildings of Scotland's past, promoting a greater appreciation of their value through the maintenance of the National Monuments Record of Scotland (NMRS) – which is now accessible online through CANMORE, and presenting them more directly by selective publications and exhibitions.
- Royal Commission on the Ancient & Historical Monuments of Wales – The Royal Commission was established in 1908 to make an inventory of the ancient and historical monuments of Wales and Monmouthshire. It is currently empowered by a Royal Warrant of 1992 to survey, record, publish and maintain a database of ancient and historical sites, structures and landscapes in Wales. It is also responsible for the National Monuments Record of Wales which is open daily for public reference, for the supply of archaeological information to the Ordnance Survey for mapping purposes and for the coordination of archaeological aerial photography.

8.11.1.1 Published Guidance in the UK

The English Heritage Building Conservation and Research Team have produced a set of bibliographical lists designed to provide basic information on classic text, and recent sources of technical information and advice relating to building conservation. These lists are intended as a useful starting place for students of conservation and professionals newly entered into the field, or for non-specialist with an interest in historic building materials and techniques. For certain subjects, particularly useful periodicals are also listed. It is intended to update and expand these lists along with additional subject areas. Technical References for the following subjects can be found here:

1 Building conservation (general)
2 Stone
3 Mortars, plasters, renders & pointing
4 Brick

More subjects are planned to follow, including terracotta, timber, metals, earth, paints, coatings, modern materials and others.

Historic Scotland produce a very wide range of technical and policy guidance on heritage and conservation – these appear as technical advice notes (TANs), as practitioner guides and as research reports. Recent titles have included:

Fire Safety Management in Heritage Buildings – which aims to reconcile the continued use of historic buildings with protection from loss by fire. Building management measures are outlined as well as an overview of legislation, standards and codes.

Corrugated Iron and Other Ferrous Metal Cladding – An outline of the history of ferrous metal cladding in Scotland. The Note seeks to reinstate the material's reputation as a significant vernacular building component. Technical guidance, documentation and conservation techniques are discussed.

The Consequences of Past Stone cleaning Intervention on Future Policy and Resources. This Research report assesses the effects different cleaning regimes had on natural stone.

There is more specific guidance on conservation techniques in British Standard BS 8221 – Code of Practice for the Cleaning and Surface Repair of Buildings – Natural Stone, Brick and Terracotta.

8.11.2 Legislative Guidance – Czech Republic

The Czech Ministry of Culture governs most of official activities including documentation, interpretation, presentation and publication of cultural heritage in the Czech Republic. The State Institute for the Care for Monuments was established to develop theoretical and methodical guidelines in the process of conservation of culture heritage and to manage *Central Directory of Listed Monuments*. Nine Regional Heritage Institutes supervise conservation and transformation activity including listed monuments and areas of protection declared by law. In the near future, this system should be simplified to only one *National Heritage Institute* which removed many offices in important historic centres.

There are three main levels of investigation and documentation activities concerning architectural heritage in the Czech Republic:

- Settlement units inventory and evaluation
- Historic monuments research and documentation
- Specialised investigation activities

The National Heritage Institute co-ordinates regulations and regeneration process in Historic Town Reserves and Urban, Rural and Landscape Areas of Protection. Together with the National Heritage Institute further institutions work in the field of identification of heritage values. Larger scale projects undertaken by Academy of Sciences include the Atlas of Historic Towns in Czech Republic and Inventory of artistic treasures in Bohemia, Moravia and Silesia.

Universities and either professional groups or associations (for example SOVAMM – *Association for Recovery of Villages and Small Towns*) join other considerable research projects, for example The Atlas of Vernacular Architecture in Czech Republic.

Presentations by Geographic Information Systems (GIS) have been successfully used for larger scale projects in Uherské Hradiště, National Park Podyjí-Thayatal, and region Ústí nad Orlicí – Vysoké Mýto. Another project has been prepared in historical town of Tel, where the methods and approaches were tested in the years of 1996–1998.

The National Heritage Institute is also responsible for field research in the Czech Republic which aims to uncover and describe the main consecutive periods of building activity during the history of each architectural monument. Using different colours investigators designate the periods into the ground plans of the building. Another part of the documentation includes particulars of the current state of monument, inventory of valuable details and list of the known historical written sources mentioning the changes. Standard methodical base for analysing of different architectural types (town houses, churches, castles) was elaborated in SÚRPMO (former *State Institute for Conservation of Historical Towns and Objects*) and has been successfully worked with there from the early 1950s to the early 1990s.

The Central State Archive in Prague collects copies of historical monuments research reports. The Journal *Historical Monuments, Research & Documentation* publishes several studies and articles regularly twice a year.

A third part of the National Heritage Institute covers all research activities dealing with single parts of bigger architectural structures, for example decoration of the facades, heating systems, timber frames or vaults. Some of them belong to the group of works, which are regulated by concessions licensed by Ministry of Culture. Appropriate documentation of the state before and after restoration is a compulsory part of the Restoration Report, which should be archived at the competent office of the Regional Heritage Institute. Historical building constructions belong to the subjects excluded from the compulsory documentation, which could be demanded by the National Heritage Institute in case of restoration works or in the case of colliding with archaeological situations. The creating of an independent system of documentation would be useful, for instance, for roof constructions. Surveying, dating and typological classification of historic timber roof frames create an important part of an architectural heritage research. The qualified care about the structures, which were put together with use of now abandoned techniques, deserves deeper insight.

8.11.3 Legislative Guidance – Spain

Article 46 of the 1978 Constitution directs the authorities to "guarantee the preservation and to promote the enrichment of the historic, cultural and artistic heritage of the peoples of Spain and of the property of which that heritage consists". The text goes beyond "conservation" to include the "enrichment" of cultural property. Acting upon this principle, the Parliament approved the *Historical Heritage Act of 1985*, a piece of legislation that broke new ground in heritage protection policy.

The dual purpose of the *1985 Historical Heritage Act* was to ensure compliance with the 1978 Constitutional mandate that enables the autonomous communities to pass their own regional laws on the same subject, as mandated by their own Charters. However, the *1985 Historical Heritage Act* was challenged by various regions on the grounds that the central government also had responsibilities in this field. The Constitutional Court, while dismissing the anti-constitutional claim, admitted that cultural properties were surely part of the national heritage of the whole country and therefore the two levels of government should work together. In practice, this meant that most communities introduced their own legislation anyway: *Basque Country (7/1990 Act); Castile-La Mancha (4/1990 Act); Andalusia (1/1991 Act); Catalonia (9/1993 Act); Galicia (8/1995 Act); Valencian Community (4/1998 Act); Madrid (10/1998 Act); Balearic Islands (Act 12/1998); Canary Islands (4/1999 Act); Extremadura (2/1999 Act); Aragon (3/1999 Act); Asturias (1/2001 Act); and Castile-Leon (12/2002 Act).*

These laws follow a more "anthropological" interpretation of cultural heritage, leaving the traditional architectural canons employed in the nineteenth and part of the twentieth centuries behind. The protective system employed by these laws is implemented via a series of administrative measures (prohibitions, fines, conservation orders, bans on sale or export, etc.), combined with incentives, such as the so-called "cultural one per cent", a levy on the cost of all public works which is used to help defray the cost of conservation. Legislation of both the central government and the regional authorities establishes various ways of defining heritage property, usually based on two categories. On the one hand, this includes the assets of cultural interest, and on the other hand, those properties included on a general inventory list of national interest. An important element of both the national and regional laws is the link made between cultural property laws and legislation for urban development.

Cultural institutions such as museums and archives are regulated by the *Historical Heritage Act of 1985*, which gives a brief definition of such bodies and the terms under which they are set up, administered and coordinated, together with how people can use their services. The *1985 Historical Heritage Act* is complemented by a series of nationwide enabling regulations governing such matters as specialist arms length institutions. It also includes a series of rules, applying to specific institutions such as the Prado Museum, the Archaeological National Museum, the Museum of America, the Museum of the Alhambra, the Reina Sofia Museum and Art Centre, the National Museum of Anthropology, the Museum of Roman Art, the National Museum of Decorative Arts, the National Museum of Ceramics and Sumptuary Arts, the National Sculpture Museum, the National Museum of Science and Technology, as well as the state-run archives (National Historical Archive, the General Archive of Simancas, the Archive of the Indies, the Archive of the Kingdom of Aragon).

As far as regional legislation is concerned, the dominant trend is to approve individual laws for museums and archives independently of national heritage legislation. Regions which have their own museum legislation include: *Andalusia (2/1984 Act), Aragon (7/1986 Act), Catalonia (17/1990 Act), Castile-Leon (10/1994 Act), Murcia (5/1996 Act), Madrid (9/1999 Act)* and *Cantabria (5/ 2001 Act)*. Regions with their own laws for public archives are: *Andalusia (3/ 1984 Act), Aragon (6/1986 Act), the Canary Islands (3/1990 Act), Murcia (6/ 1990 Act), Castile-Leon (6/1991 Act), Madrid (Act 4/1993 Act), La Rioja (4/ 1994 Act), Catalonia (13/2001 Act)* and *Cantabria (3/2002 Act)*.

8.11.4 Responding to Threats to the Cultural Heritage

A survey by the University of Loughborough (Matthews, 2006) reported that the threat to the moveable cultural heritage came from:

- Water related disasters – 68%
- Fire – 11%
- Vandalism – 3%
- Theft/break-in/burglary – 6%,
- Bomb incident/terrorist threat – 2%
- Other (e.g. building collapse, IT, power failure, etc.) – 10%

Since this survey concerned moveable heritage it is not surprising that air pollution is not included (see below for further discussion on possible impacts of air pollution on indoor objects). However, it is interesting to note that all of the threats are those which occur suddenly, and often very dramatically. It does not include slow and more progress effects which are less apparent to the visitor and also which are more likely to attract the attention of the media.

The same survey concluded that the most frequent lessons learned from experiencing a disaster, in order, were:

- Awareness of building issues, e.g. maintenance, regular checks, updating plans;
- The importance of training, particularly with regard to knowledge of the disaster control plan and how to respond in a disaster;
- Availability of adequate emergency equipment in appropriate locations.

It is interesting to consider that although these findings are based on responses from those concerned with moveable heritage similar lessons could be applied equally well to the immovable built cultural heritage – particularly the need for awareness of the threats and training in how to respond to the threat.

8.12 Conclusions

Management of heritage at the local level is clearly complex and can require both reactive and proactive interventions. This chapter has presented a number of situations and examples that demonstrate ways that the more theoretical sections of the book may be applied to inform local policy making and management strategies. These can be seen as part of both management of the risk from pollution and the response to regulatory and statutory requirements.

Acknowledgements The authors are grateful to all the members of CULT-STRAT and other co-operative projects for their help with this chapter. We are grateful to Petr Měchura, ITAM and Petr Chotěbor from the Czech President's Office.

References

Bacci, M., Cucci, C., Dupont, A.-L., Lavédrine, B., Picollo, M. and Porcinai, S. (2003) Disposable indicators for monitoring lighting conditions in museums. Environmental Science and Technology, 37, 5687–5694.

Blades, N., Oreszczyn, T., Bordass B. and Cassar, M. (2000) Guidelines on pollution control in museum buildings. Published by the Museums Association and distributed with MUSEUM PRACTICE, 15, Nov. London.

Blades, N., Grøntoft, T., Dahlin, E., Taylor, J. and Rentmeister, S. (2006) Chap. 2.2.3. Advantages of dosimetry as an environmental monitoring strategy. In Dahlin, E. (Ed.) EU project MASTER (EVK-CT-2002-00093) Publishable Final Report "The MASTER Early Warning System and Preventive Conservation Strategy".

Brimblecombe, P., Blades, N., Camuffo, D., Sturado, G., Valentino, A., Gysels, K., Van Grieten, R., Busse, H.-J., Kim, O., Ulrych, U. and Wieser, M. (1999) The Indoor Environment of a Modern Museum Building, the Sainsbury Centre for Visual Arts, Norwich, UK. Indoor Air, 9, 146–164.

Brimblecombe, P. (1990) The composition of museum atmospheres. Atmospheric Environment, 24B, 1–8.

Brimblecombe, P., (1988) The composition of museum atmospheres. Atmospheric Environment 24B(1), pp. 1–8.

Brimblecombe, P. (2006). Composition and Chemistry of Museum Air. Plenary Presentation to the 7th Indoor Air Quality 2006 Meeting (IAQ2006). Braunschweig, Germany, 15–17 November 2006.

Bullock, L. and Saunders, D. (1999) Measurement of cumulative exposure using Blue Wool standards. In: ICOM-CC 12th Triennial, Lyon. Ed. by: J. Bridgland. London, James and James. pp. 21–26.

Camuffo, D., Van Grieken, R., Busse, H.-J., Sturaro, G., Valentino, A., Bernardi, A., Blades, N., Shooter, D., Gysels, K., Deutsch, F., Wieser, M., Kim, O. and Ulrych, U. (2001) Environmental monitoring in four European museums. Atmospheric Environment, 35, Supplement No. 1, pp. S127–S140.

Cano-Ruiz, J.A., Kong, D., Balas, R.B., Nazaroff, W.W. (1993) Removal of reactive gases at indoor surfaces. Combining mass transport and surface kinetics. Atmospheric Environment, 27 A(13): 2039–2050.

Cass, R.G., Nazaroff, W.W., Tiller, C. and Whitmore, P.M. (1991) Protection of works of art from damage due to atmospheric ozone. Atmospheric Environment, 25A, 441–451.

CEN/CT346 (2006) www.cenorm.be/cenorm/index.htm

Chaloulakou, A. and Mavroidis, I. (2002) Comparison of indoor and outdoor concentrations of CO at a public school. Evaluation of an indoor air quality model. Atmospheric Environment, 36, 1769–1781.

Chen, F., Yu, S.C.M. and Lai, A.C.K. (2006) Modelling particle distribution and deposition in indoor environments with a new drift–flux model. Atmospheric Environment, 40, 357–367.

Czop, J. (2002). Air Pollution Deposition in Museums. Proceedings of the 5th EC Conference, May 16–18 2002, Krakow Poland.

Dahlin, E. (Ed.). (2006) EU project MASTER (EVK-CT-2002-00093) Publishable Final Report "The MASTER Early Warning System and Preventive Conservation Strategy".

Dahlin, E., Grøntoft, T., Rentmeister, S, Calnan, C., Czop, J., Hallet, K., Howell, D., Pitzen, C. and Sommer Larsen, A. (2005) Development of an early warning sensor for assessing deterioration of organic materials indoor in museums, historic buildings and archives. In: 14th Triennial Meeting The Hague, 12–16 September 2005, Preprints, Volume II, 617–624.

Gegisian, I. (2007) Assessing the contribution of local air quality management to environmental justice in Eng-land and Wales. Faculty of Applied Sciences, University of the West of England, Bristol.

Dimitropoulou, C., Ashmore, M.R., Byrne, M.A. and Kinnersley, R.P. (2001) Modelling of indoor exposure to nitrogen dioxide in UK. Atmospheric Environment, 35, 269–279.

Druzik, J.R., Adams, M.S., Tiller, C. and Cass, G.R. (1990) The measurement and model predictions of indoor ozone concentrations in museums. Atmospheric Environment, 24A, 1813–1823.

Ekberg, L.E. (1994) Outdoor Air Contaminants and Indoor Air Quality under Transient Conditions. Indoor Air, 4, 189–196.

Fan, Y. (1995) CFD modelling of the air and contaminant distribution in rooms. Energy and Buildings, 23, 33–39.

Fuchs, D.R., Roemich, H. and Schmidt, H. (1991) Glass-Sensors: Assessment of complex corrosive stresses in conservation research, Materials Research Society symposia proceedings, 185, 239–251.

Grøntoft, T. (2004) Deposition of Gaseous pollutants to indoor material surfaces. Time, humidity and temperature dependent deposition of O3, NO2 and SO2. Measurement and modelling. Thesis for the degree of dr. scient. Department of Chemistry. Series of dissertations submitted to the Faculty of Mathematics and Natural Sciences. University of Oslo. No. 324. Unipub AS, Oslo.

Grøntoft, T., Henriksen J. F., Dahlin, E., Lazaridis, M., Czop, J., Sommer-Larsen, A., Hallett, K., Calnan, C., Pitzen, C. and Cassar, J.A. (2005 a) Sensor and environmental data from the field test programme. EU project MASTER (EVK-CT-2002-00093) Deliverable no. D. 3.1, WP 3. Restricted data file report to the European Commission.

Grøntoft, T. and Raychaudhuri M. R. (2004) Compilation of tables of surface deposition velocities for O3, NO2 and SO2 to a range of indoor surfaces. Atmospheric Environment, 38(4), 533–544.

Gysels, K., Delalieux, F., Deutsch, F., Van Grieken, R., Camuffo, D., Bernardi, A., Sturaro, G., Busee, H-J. and Wieser, M., (2004) Indoor environment and conservation in the Royal Museum of Fine Arts, Antwerp, Belgium. Journal of Cultural Heritage, 5, 221–230.

Hayes, S.R. (1989) Estimating the effect of being indoors on total personal exposure to outdoor air pollution. Journal of the Air Pollution Control Association, 39, 1453–1461.

Hayes, S.R. (1991) Use of an indoor air quality model (IAQM) to estimate indoor ozone levels. Journal of the Air & Waste Management Association, 41, 161–170.

IMPACT, 2006. EU project IMPACT EVK-CT-2000-00031. Innovative Modelling of Museum Pollution and Conservation Thresholds. www.ucl.ac.uk/sustainableheritage/impact

ISO 11 844 Corrosion of metals and alloys — Classification of low corrosivity of indoor atmospheres — Part 1: Determination and estimation of indoor corrosivity.

Jones, A.P. (1999) Indoor air quality and health. Atmospheric Environment, 33, 4535–4564.

Kucera, V. (2005) EU project MULTI-ASSESS. Model for multi pollutant impact and assessment of threshold levels for cultural heritage. EVK4-CT-2001-00044. Final Publishable Report.

Larsen, R. (Ed.) (1996) Deterioration and conservation of vegetable tanned leather, European Commission, Environment Leather Project, Research report no. 6.

Lazaridis M., Glytsos, T., Aleksandropoulou V. and Kopanakis I. (2006) Chap. 2.2.4, Indoor/outdoor modelling. In Dahlin, E. (Ed.) EU project MASTER (EVK-CT-2002-00093) Publishable Final Report "The MASTER Early Warning System and Preventive Conservation Strategy".

Lazaridis, M., Glytsos, T., Grøntoft, T., Blades, N., Aleksandropoulou, V. and Kopanakis, I. (2006) Chap. 2.5. The use of Indoor/outdoor modelling for cultural heritage sites. In Dahlin, E. (Ed.) EU project MASTER (EVK-CT-2002-00093) Publishable Final Report "The MASTER Early Warning System and Preventive Conservation Strategy".

Leißner, J. and Fuchs, D.R. (1992) Glass sensors: A European study to estimate the effectiveness of protective glazing at different cathedrals. Proceedings of Congreso Internacional de Rehabilitacion del Patrimonio Arquitectonico y Edificacion, Islas Canarias, 285–290.

Liu, De-Ling and Nazaroff, W.W. (2001) Modelling pollutant penetration across building envelopes. Atmospheric Environment, 35, 4451–4462.

MASTER (2006) (EVK-CT-2002-00093) www.nilu.no/master/

Matthews, G. (2006). Safeguarding heritage at risk: disaster management in United Kingdom archives, libraries and museums. UK survey results – summary. www.lboro.ac.uk/departments/dis/disresearch/heritage%20 at%20risk/Uksurveyresults.pdf Validated 31/08/07.

Mitchell G and Dorling D (2003) An environmental justice analysis of British air quality. Environment and Planning A, 35, 909–929.

Muller, C., Corel, R. and van Dijke, R. (2006). Air Quality Monitoring in European Museums 2000 – present. Plenary Presentation to the 7th Indoor Air Quality 2006 Meeting (IAQ2006). Braunschweig, Germany, 15–17 November 2006.

Nazaroff, W.W. and Liu, De-Ling (2001) Modelling pollutant penetration across building envelopes. Atmospheric Environment, 35, 4451–4462.

Nazaroff, W.W. and Cass, G.R. (1986) Mathematical modelling of chemically reactive pollutants to indoor air. Environmental Science and Technology, 20, 924–934.

NSCA (2007) http://www.nsca.org.uk/pages/policy_areas/air_quality_guidance.cfm

Oddy, W.A. (1973) An unsuspected danger in display. Museums Journal, 73, 27–28.

Odlyha, M., Wade, N., Wang, Q. et al., (2005) "Microclimate Monitoring: damage assessment for cultural heritage preservation " Preprints Vol. 2 (p. 670–677) 14th Triennial Meeting The Hague 12–16 September 2005.

Papakonstantinou, K.A., Kiranoudis, C.T. and Markatos, N.C. (2000) Mathematical modelling of environmental conditions inside historical buildings. The case of the archaeological museum of Athens. Energy and Buildings 31, 211–220.

Sebera, D. K. (1994) Isoperms: An environmental Management Tool. Commission of Preservation and Access, Washington DC.

Salmon, L.G., Cass, G.R. and Bruckman, J.H. (2000) Ozone exposure in the historic central district of Krakow, Poland (2000). Atmospheric Environment, 34, 3823–3832.

Spengler, J.D. and Sexton, K. (1983) Indoor air pollution: a public health perspective. Science, 221, 9–17.

Taylor, J., Blades, N., Cassar, M., Grøntoft, T., Dahlin, E., Rentmeister, S., Hanko, H. and Heinze, J. (2006) Chap. 2.7. Preventive Conservation Strategy. In: Dahlin, E. (Ed.) EU project MASTER (EVK-CT-2002-00093) Publishable Final Report "The MASTER Early Warning System and Preventive Conservation Strategy".

Tetreault, J. (2003) Airborne Pollutants in Museums, Galleries, and Archives: Risk Assessment, Control Strategies, and Preservation Management. Canadian Conservation Institute, Ottawa.

Thatcher, T.L. and Layton, D.W. (1994) Deposition, resuspension, and penetration of particles within a residence. Atmospheric Environment, 29, 1487–1497.

Weschler, S. J. and Shields, H.C. (1997) Potential reactions among indoor pollutants. Atmospheric Environment, 31(21): 3487–3495.

Zinn, E., Reilly, J. M., Adelstein, P. Z. and Nishimura, D. W. (1994) Air pollution effects on library microforms. In: Preventive Conservation Practice, Theory and Research. Preprints on the Contributions to the Ottawa Congress, 12–16 September 1994. Ed. by: Roy, A. and Smith, P. London, IIC, pp. 195–201.

Sources of Additional Information

Further information on the Rapid Assessment kits for damage assessment and their availability can be found on the homepage of the CULT-STRAT project at http://www.swereakimab.se/web/page.aspx?pageid = 8529

IAQ, Indoor Air Quality in museums and archives, http://iaq.dk/index.html – index
(This page gives links to other indoor air quality and conservation resources)

IMPACT homepage, http://www.ucl.ac.uk/sustainableheritage/impact/index.htm

NIST Multizone Modelling Website, (www.bfrl.nist.gov/IAQanalysis/)

CHAM – PHOENICS homepage. Pioneering CFD Software for Education & Industry, http://www.cham.co.uk/

EMITS Project ("Environmental Monitoring of Integrated Transport Strategies- (EMITS)", Oxford – United Kingdom LIFE95 ENV/UK/000595).

Winkler, E.M. (1994) Stone in architecture – properties, durability. Springer Verlag, Berlin Heidelberg.

Chapter 9
Air Quality Policy

James Irwin, Johan Tidblad, and Vladimir Kucera

9.1 Overview

There are two main ways to protect cultural heritage objects from air pollution damage. First, by reducing ambient levels of air pollution and second, by instigating local management strategies that either reduce the impact of pollution or repair the damage.

Air pollutants are major contributors to the deterioration of many materials used in cultural heritage objects, and are emitted throughout the World by industrial activities and the transport sector. These pollutants create problems on a local scale but are also transported over long distances. Hence, both regional and local policy interventions have a role to play in reducing concentrations. International action in the form of European Union Directives and protocols of the United Nations Economic Commission for Europe are addressing adverse effects on human health and ecosystems. These initiatives, however, have not, as yet, taken specific account of effects on building materials although such materials may be sensitive to pollution at even lower levels than biological systems.

Europe has an enormous stock of cultural heritage objects that cost billions of euros to maintain. Moreover, damage caused to these objects endangers an important part of our cultural identity. There is an urgent need, therefore, to ensure that the impacts of pollution on monuments and buildings are considered in policy making on air quality, in addition to the human health and ecosystem damage that currently dominate the debate.

After some brief reflections on the process of policy development and characteristics of good policy, this chapter considers how research on air pollution, corrosion, soiling and their economic impact may be used to help to develop integrated air quality policy. As heritage can deteriorate even in the absence of pollution it is necessary to define a tolerable level of deterioration, and thereby tolerable (sometimes called acceptable) pollution levels, to aid policy formulation.

J. Irwin (✉)
University of the West of England, Coldharbour Lane, Bristol, BS16 1QY, UK
e-mail: jimiirwin@aol.com

J. Watt et al. (eds.), *The Effects of Air Pollution on Cultural Heritage*,
DOI 10.1007/978-0-387-84893-8_9, © Springer Science+Business Media, LLC 2009

One approach to addressing this question is described. A consideration of ways in which this information can inform policy development at both national and local levels follows, including some factors to consider in developing local air quality management options.

9.2 Factors in Policy Development

At its simplest a policy can be thought of as a statement of intent to achieve certain objectives and outcomes. While it states what is to be achieved it does not necessarily define how the outcomes will be realised or describe the mechanisms that could be used. Hence a policy objective could be 'to ensure that no object of cultural heritage suffers damage from air pollution' but the real challenge lies in developing and implementing mechanisms to deliver it.

New policy can become necessary if there is no existing policy or when existing policy is proving insufficient. Drivers for new or revised policy include:

- Operational experience
- A changing situation
- New scientific evidence
- Changing priorities

Experience may show that the desired objectives are not being attained. Cultural objects may be continuing to corrode or become soiled, or the rate of improvement may be less than anticipated. In assessing this type of evidence it is important to consider spatial variability and establish whether any apparent shortcoming is localised or widespread. Nor does continuing observation of an effect necessarily imply that the causal relationship is unchanged as, for example, when a changing pollution climate results in continuing corrosion but due to a different mix of pollutants.

The pollution climate of Europe has changed significantly over the last fifty years. Initially, following the smogs of the 1950s, controls on coal burning in urban areas led to sharp reductions in black smoke concentrations as illustrated for London in Fig. 9.1. Subsequent controls on emissions from power stations, refineries and other stationary sources have been accompanied by an increase in vehicle registrations and kilometres driven. These, and other, changes have created a new multi-pollutant situation where SO_2 is no longer the dominating corrosive pollutant, as discussed in Chapter 1. It has, therefore, been necessary to derive and utilise new dose-response functions to address deterioration in a multi-pollutant environment, and to consider mixtures of pollutants when developing abatement strategies.

New scientific evidence may show that current and likely future levels of pollutants will be sufficiently high to cause damage. Some materials may be more sensitive to air pollutants than previously believed. And the damage from

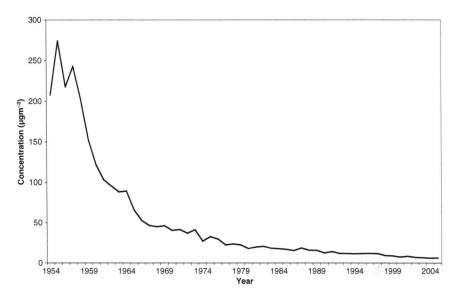

Fig. 9.1 Annual mean black smoke concentrations in London 1954–2005. (after Stedman and Maynard, 2007)

combinations of pollutants may be more severe than for the pollutants acting singly.

In the case of the impacts of air pollution on our cultural heritage there is a strong scientific base on which to draw. Work to understand the damage caused by air pollution to materials and cultural heritage began in the late 1970s. In the United States of America, Congress authorised the National Acid Precipitation Assessment Programme (NAPAP), with the objective of providing a sound scientific and technical understanding of inter-state air pollution and acid rain, as a basis for possible revision of the Clean Air Acts. Within this initiative the NAPAP materials research programme was designed as an integrated study to investigate the incremental effects of wet and dry deposition, characterised by chemical species, on materials durability.

In Europe, the United Nations Economic Commission for Europe (UNECE) launched an International Cooperative Programme to investigate the effects of air pollutants on materials, including historic and cultural monuments (ICP Materials). This research programme is an integral component of the Convention on Long-Range Transboundary Air Pollution. Fundamental research on the processes involved in cultural heritage damage, including the role of air pollution, was also investigated in a series of major projects funded by the European Commission under the RTD Framework Programmes (CORDIS 2007).

Finally public priorities may change. The public may attach more importance to its cultural heritage than it once did. This can be for a number of reasons – concern at the loss of heritage meaning more importance is attached to what remains, generally increased awareness of our environment and an increasing acceptance of the need for behaviour change at an individual level.

9.3 Elements of Good Policy

To date air pollution policy in Europe has focused on the protection of human health, the natural environment and, to a lesser extent, agricultural crops. But whatever the area of concern there are a number of characteristics of good policy:

- Evidence based
- Forward looking
- Innovative
- Flexible
- Learns lessons
- Inclusive
- Joined-up
- On-going review and evaluation

In considering evidence it is important to consider not only scientific evidence as summarised in this volume but also social evidence – how do people feel about an issue? The importance of considering public perception has been evident in current debates about genetically modified crops and nanotechnology (Wilsdon and Willis 2004). In both cases public concern centres largely on possible risks to health although there are other questions including the direction of travel of a new technology ... will it benefit all society equally, or at all? Traditional handling of risk management in many policy areas was characterised by assertions of safety, denial of uncertainty, informing the public and narrow framing of decisions. The ability to accept and acknowledge uncertainty is a key component of new ways of handling risk which also emphasise public involvement and broad framing of decisions. While questions of risk are often uppermost in the debate they can reflect other questions about visions and values. In analysing the impacts of air pollution it is, therefore, helpful to consider not only technical characteristics of the harm that can arise, including spatial and temporal extent, severity and irreversibility, but also social aspects that influence public response. These include dread, distrust, equity and imposition (Irwin et al., 2002).

Policy should be forward looking, taking account of possible future changes. For example, air pollution and anthropogenic climate change are closely linked in several ways. Both arise from burning fossil fuels: sulphur and nitrogen oxides cause air pollution and carbon dioxide contributes to climate change. In some cases there may be significant synergies and joint benefits through co-ordination of climate change and local air quality policies (Williams 2007). In particular, burning smaller amounts of fossil fuels would result in a fall in SO_2, NO_x and fine particles, leading to a saving in the costs of tackling air pollution, and reducing the number of premature deaths. Addressing this type of issue is a good example of joined-up policy. Another, local scale, example would be the pedestrianisation of a town centre thereby reducing pollution, improving road safety and contributing to public health by encouraging walking.

Finally it is essential that policy and mechanisms are regularly reviewed and evaluated. Figure 9.2 shows NO_x emissions from motor vehicles. The relative contributions of four policy initiatives can be seen: a shift from petrol to diesel, the introduction of 3-way catalysts, emission standards for heavy duty vehicles and emission standards for passenger cars. Over the period from 1987 to 1998 it is clearly 3-way catalysts that have made the greatest contribution, reducing emissions from over 7 Mtonnes in the reference scenario (if there had been no policy intervention) to well under 5 Mtonnes, with smaller contributions from the other measures (EEA 2007a).

When evaluating a policy it is also important to look forward and understand the extent to which a current initiative will continue to deliver further improvement. Figure 9.3 illustrates how improved fuel efficiency has partly offset increased kilometres driven over the last 15 years (EEA 2007b). An estimate of the extent to which this can continue is an important element in formulating policy to reduce NO_2 concentrations.

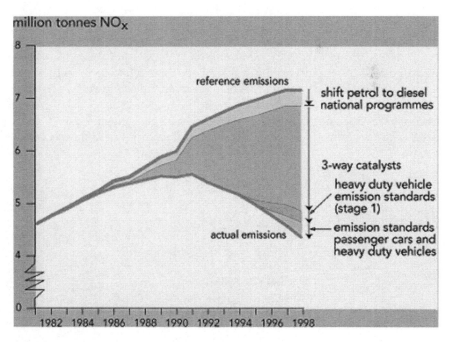

Fig. 9.2 Contributions of policy initiatives to decreasing NOx emissions from vehicles in Europe. (© EEA 2007a)

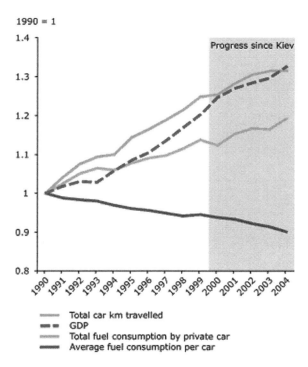

Fig. 9.3 Growth in private car travel versus fuel efficiency in EU 15. (© EEA 2007b)

But when all the scientific evidence, both technical and social, is considered a policy is essentially a political decision. Figure 9.4 shows an idealised relationship between corrosion rate and pollutant concentration. In the case of cultural heritage the concept of a threshold concentration such as a critical load or level does not apply as any amount of pollution can lead to some deterioration. Hence it becomes necessary to agree a tolerable/acceptable

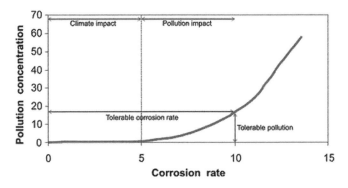

Fig. 9.4 Idealised relationship between corrosion rate and pollutant concentration

rate of deterioration. Do we aim to limit corrosion to the climate impact? Or tolerate a corrosion rate defined as a multiple of the background rate of deterioration etc? Approaches to addressing this question are described later. But inevitably this type of decision is informed by the costs involved to reduce concentrations and the value of the benefits obtained, as discussed in Chapter 7.

9.4 The Regulatory Regime in Europe

Exposure to air pollutants in ambient air can have serious effects on human health. High pollution episodes lead to increased hospital admissions and contribute to premature death of more vulnerable sections of the community. In terms of lives lost the impacts are even higher than those of car accidents (EEA 2007b). Air pollution also has negative effects on the environment both through direct effects of pollutants on vegetation and cultural heritage, and indirectly through effects on the acid and nutrient status of soils and waters (NEGTAP 2001).

These effects have been recognised for a long time and there has been a great amount of work to establish 'safe' concentrations below which pollutants do not have adverse effects. This research has covered human health, natural ecosystems, crops, building materials and, as discussed in this volume, our cultural heritage. In some cases it has been established that there is no safe level. In these cases policy now focuses on an 'exposure reduction' approach, described later.

9.4.1 European Air Quality Standards and Guidelines

The European Union (EU) has been legislating to control emissions of air pollutants and to establish objectives for ambient air quality since the 1980s (EU 2007). Ambient air quality legislation was consolidated with the adoption in 1996 of the Air Quality Framework Directive (AQFD). This established a strategic framework for tackling air quality by setting European-wide limit or target values for twelve pollutants in a series of Daughter Directives. The twelve pollutants are sulphur dioxide, nitrogen dioxide, particulate matter, lead, carbon monoxide and benzene, for which limit values have been set; and ozone, polycyclic aromatic hydrocarbons, cadmium, arsenic, nickel and mercury, which should be reduced to levels as low as practicable. Limit values are legally binding from the date they enter into force and must not be exceeded, subject to any exceedances permitted in the legislation. They are set for individual pollutants and include a concentration, an averaging time over which it is to be measured and, for some pollutants, a number of exceedances that can be permitted in a year. For some pollutants, such as SO_2 there is more than one limit value covering different endpoints or averaging times because the observed health impacts occur over different exposure times. In addition to limit values some EU Directives include 'target values'. These are similar to limit values but are not legally binding and should be attained where possible by taking all possible measures not entailing disproportionate costs. In addition, the Daughter Directives define alert thresholds, and specify monitoring strategies, measuring methods, quality assurance procedures, and modelling criteria for each pollutant. Examples of current EU limit and target values are listed in Table 9.1 (Europa 2007).

A new Ambient Air Quality Framework Directive was adopted in 2008. This will streamline existing Directives and introduce flexibilities in meeting obligations under some circumstances. Significantly, in line with advice from

Table 9.1 EU limit and target values for the protection of human health

Pollutant	Concentration	Averaging period	Legal status	Permitted exceedances each year
Sulphur dioxide (SO$_2$)	350 μg m^{-3}	1 hour	Limit value entered into force 1.1.2005	24
	125 μg m^{-3}	24 hours	Limit value entered into force 1.1.2005	3
Nitrogen dioxide (NO$_2$)	200 μg m^{-3}	1 hour	Limit value entered into force 1.1.2010	18
	40 μg m^{-3}	1 year	Limit value entered into force 1.1.2010	n/a
PM$_{10}$	50 μg m^{-3}	24 hours	Limit value entered into force 1.1.2005	35
	40 μg m^{-3}	1 year	Limit value entered into force 1.1.2005	n/a
PM$_{2.5}$	25 μg m^{-3}	1 year	Limit value enters into force 1.1.2015	n/a
Lead (Pb)	0.5 μg m^{-3}	1 year	Limit value entered into force 1.1.2005 (or 1.1.2010 in the immediate vicinity of specific, notified industrial sources)	n/a

Table 9.1 (continued)

Pollutant	Concentration	Averaging period	Legal status	Permitted exceedances each year
Carbon monoxide (CO)	$10\ mg\,m^{-3}$	Maximum daily 8 hour mean	Limit value entered into force 1.1.2005	n/a
Benzene	$5\ \mu g\,m^{-3}$	1 year	Limit value enters into force 1.1.2010	n/a
Ozone	$120\ \mu g\,m^{-3}$	Maximum daily 8 hour mean	Target value enters into force 1.1.2010	25 days averaged over 3 years
Arsenic (As)	$6\ ng\,m^{-3}$	1 year	Target value enters into force 1.1.2012	n/a
Cadmium (Cd)	$5\ ng\,m^{-3}$	1 year	Target value enters into force 1.1.2012	n/a
Nickel (Ni)	$20\ ng\,m^{-3}$	1 year	Target value enters into force 1.1.2012	n/a
Polycyclic Aromatic Hydrocarbons	$1\ ng\,m^{-3}$ (expressed as concentration of benzo(a)pyrene)	1 year	Target value enters into force 1.1.2012	n/a

the World Health Organisation, the Directive introduces controls on fine particles, $PM_{2.5}$ (WHO 2006). As this is believed to be a 'non-threshold' pollutant, an 'exposure reduction' approach has been introduced in addition to a target value. This aims to improve air quality where the greatest number of people are likely to be exposed, rather than pollution 'hotspots'. This will require reductions in 3-year running mean concentrations averaged over selected monitoring stations in urban background areas. The reduction to be achieved by 2020 will range from 0 to 20% depending on concentrations in the area in 2010.

The EU has also set target values for the protection of vegetation and natural ecosystems. In the case of SO_2 these incorporate both an annual and a winter mean. For O_3 the situation is a little more complex as it is believed that a critical factor is cumulative exposure during the growing season. This has been addressed by a target value for exposure, assessed as accumulated one-hour ozone concentrations above a threshold of 40 ppb – AOT40. In 2004 this was exceeded in over a quarter of the agricultural area of the 32 EEA countries (EEA 2007b). Current EU target values for the protection of vegetation and ecosystems are shown in Table 9.2.

Table 9.2 EU limit and target values for the protection of vegetation and ecosystems

Pollutant	Concentration	Averaging period
Nitrogen oxides (as NO_2)	$30\ \mu g\,m^{-3}$	Annual mean
Sulphur dioxide	$20\ \mu g\,m^{-3}$	Annual and winter mean
Ozone	$18\ \mu g\,m^{-3}$	based on AOT40 for May to July.

9.4.2 National Air Quality Objectives

Within the European Union individual countries can also define their own objectives. For example, in the United Kingdom some national and regional air quality objectives have been set. Standards have been derived as benchmarks for setting objectives. These are set as minimum or zero risk levels, based solely on scientific and medical evidence as to the effects of particular pollutants on health or the wider environment (COMEAP 2007). This is similar to the approach adopted by the World Health Organisation in the formulation of their air quality guidelines (WHO 2006).

In moving from these standards to objectives the UK government has taken account of economic efficiency, practicability, technical feasibility and time-scales. These objectives do not have legal force but their existence and attainment has to be borne in mind when designing and executing all measures to improve air quality. Examples of where these national and regional objectives are more stringent than EU legislation are summarised in Table 9.3. In the case of $PM_{2.5}$ there is an objective to reduce average concentrations in urban background areas by 15% but with a backstop objective that nowhere should concentrations exceed 25 ugm^{-3} (Defra 2007).

While these examples aim to improve human health, national or regional objectives could be set to protect cultural heritage in areas with significant stock-at-risk, as described in Chapter 6.

Table 9.3 EU limit values and UK air quality objectives

Pollutant	EU legislation	UK objective	Scotland
PM_{10} annual mean	$40\ \mu g\,m^{-3}$	$40\ \mu g\,m^{-3}$	$18\ \mu g\,m^{-3}$
$PM_{2.5}$ exposure reduction	% reduction determined by AEI in 2010 (ii)	15% exposure reduction in urban background	$12\ \mu g\,m^{-3}$
Benzene annual mean	$5\ \mu g\,m^{-3}$	$5\ \mu g\,m^{-3}$	$3.25\ \mu g\,m^{-3}$ (i)
Polycyclic Aromatic Hydrocarbons B[a]P	$1\ ng\,m^{-3}$	$0.25\ ng\,m^{-3}$	

(i) Running annual mean, also applies in Northern Ireland
(ii) AEI = average exposure indicator, a 3-year running annual mean of concentrations at urban background locations.

9.5 The Regulatory Regime in the United States of America

In the United States of America the Clean Air Act requires the Environmental Protection Agency to set national ambient air quality standards for pollutants considered harmful to public health and the environment. The act established two types of air quality standards. Primary standards set limits to protect public health, including the health of sensitive populations such as asthmatics, children and the elderly. Secondary standards set limits to protect public welfare, including protection against decreased visibility, damage to animals, crops, vegetation and buildings (EPA 2007a). Standards have been set for six principal pollutants: CO, Pb, NO_2, particulate matter as PM_{10} and $PM_{2.5}$, O_3 and SO_2 (EPA 2007b).

9.6 Critical Loads and Levels

The history of the Convention on Long-Range Transboundary Air Pollution (LRTAP) goes back to the early 1960s, when the connection between emissions of sulphur pollutants in Europe and the acidification of lakes in Scandinavia was first recognised. During the 1970s an increasing body of evidence established that pollutants can be transported hundreds of kilometres from their point of emission and affect air quality and ecosystem health. In 1979 the Convention was adopted as the first multilateral treaty aiming to protect the environment from the growing threat of acid deposition. To ensure that the Convention was underpinned by sound science:

- A network of monitoring sites throughout the ECE region was created within The Cooperative Programme for Monitoring and Evaluation of the Long-range Transmission of Air Pollutants in Europe (EMEP).
- A Working Group on Effects was established to address the effects of sulphur compounds and other major air pollutants on human health and the environment.

Initially discussions focussed on damage to forests and freshwaters but also considered detrimental effects on materials and human health. International Cooperative Programmes (ICP) were established to carry out more detailed studies and monitor long-term effects on ecosystems and materials. Today there are six ICPs, one of which considers effects on materials, including historic and cultural monuments (ICP Materials) as described in Chapter 3.

The Convention has provided a flexible framework under which 51 Parties have cooperated to improve scientific understanding of the problems. This has led to legally binding protocols covering those air pollutants with the greatest impact on the environment and human health. The first 1985 Sulphur Protocol adopted a flat-rate approach of a 30% reduction in annual sulphur emissions by all Parties. Since then there have been a series of protocols setting increasingly stringent emission reduction targets. More recently, the early single pollutant protocols have been superseded by a more complex multi-effect, multi-pollutant instrument that attempts to take account of the links between pollutant emissions and effects. This 1999 Gothenburg protocol considered the effects of SO_2, NO_x, VOCs and NH_3 on acidification, eutrophication and of ground level O_3 on people and vegetation, but did not take direct account of effects on materials (UNECE 1999).

Convention objectives include critical loads and critical levels for the protection of (semi) natural ecosystems. Critical loads and levels are usually defined as 'a quantitative estimate of exposure to one or more pollutants below which significant harmful effects on specified sensitive elements of the environment do not occur according to present knowledge' (Nilsson and Grennfelt 1988). A critical load relates to the quantity of pollutant deposited from air to the ground (as either wet or dry deposition or cloud water interception), whereas a critical level is the concentration of a gaseous pollutant in the air. There are a number of approaches

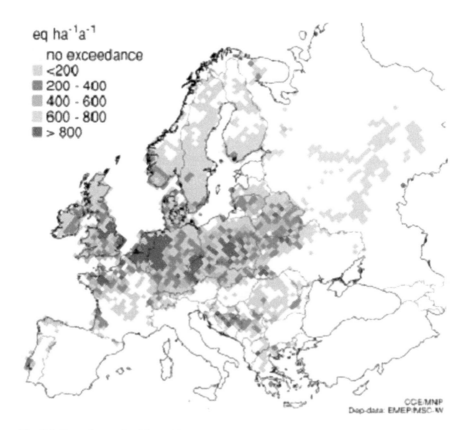

Fig. 9.5 Exceedance of acidity critical loads over Europe 2000. (CCE 2005)

to calculate and map critical loads. These range from using existing data to classify ecosystems into different levels of sensitivity, to the use of dynamic models which allow consideration of temporal aspects of exceedance and recovery.

Although the situation is improving, acidifying deposition is still above critical loads in parts of central and north-western Europe. For example, across the EU-25 approximately 15% of (semi) natural ecosystem areas were subject to acid deposition beyond their critical load in 2004 (EEA 2007c), while the percentage of EU-25 forest areas receiving acid deposition above their critical load is projected to decrease from 23% in 2000 to 13% in 2020. For those areas still at risk, ammonia is projected to be the dominant source of acidification (EEA 2007b). Eutrophication – excess nitrogen deposition – also poses a threat to a wide range of ecosystems, endangering biodiversity through changes to plant communities. Excess nitrogen deposition is widespread and expected to improve only slightly under current legislation, mainly due to the relatively small projected decline in ammonia emissions. Figure 9.5 shows critical loads exceedances for acidity in 2000 (CCE 2005).

9.7 Local Air Quality Management

While many elements of air quality policy and management are best delivered at the level of the nation state, actions at a local level can be more effective and efficient in delivering improvements at air quality 'hotspots'. This is in line with the principle of subsidiarity.

The U.S. Clean Air Act is a Federal Act covering the whole country. But complex relationships and divisions of responsibility exist between federal, state, tribal and local agencies. State and local governments are given much of the responsibility to carry out the provisions of the Act, including the development of strategies and plans, on the basis that pollution control often benefits from an understanding of local factors (Baldauf et al., 2004). States have to develop State Implementation Plans (SIPs) that outline how the state will control air pollution under the Clean Air Act. The Tribal Authority Rule recognises that tribes may develop and implement plans specific for implementation in Indian Country. The individual states or tribes may have stronger air pollution laws, but they may not have weaker pollution limits than those set by the U.S. EPA.

European legislation on air quality is built on certain principles. The first of these is that the Member States divide their territory into a number of zones and agglomerations. In these zones[1] and agglomerations,[2] the Member States should undertake assessments of air pollution levels using measurements and

---

[1] Defined as part of the territory designated by Member States.

[2] Defined as areas with a population exceeding 250,000 inhabitants or less populated areas characterised by high population density, such as more than 1,000 inhabitants/km^2.

modelling and other empirical techniques. The AQFD requires Member States
to assess air quality, introduce action plans if pollution levels are above the limit
values, and to maintain good air quality. In addition, information on air quality
should be disseminated to the public.

The Directive also requires Member States to designate a government body
to be responsible for the implementation of the Directive. This body assesses air
quality in each agglomeration and zone, and compares it with the limit values
using common methods and criteria.

The AQFD also introduced the concept of a margin of tolerance, which
allows Member States to develop an air quality improvement plan to ensure
that compliance with limit values can be achieved within the attainment dates.
Where pollution levels are above the margin of tolerance, an action plan must
be developed within two years outlining the policies and measures so that the
limit value will be attained by the target date. If the levels are between the limit
value and margin of tolerance, a Member State must submit an annual report
to the Commission. If pollution levels are below the limit value, good air
quality needs to be maintained and a report must be submitted every three
years.

Figure 9.6 illustrates the framework for managing air quality according to
the AQFD.

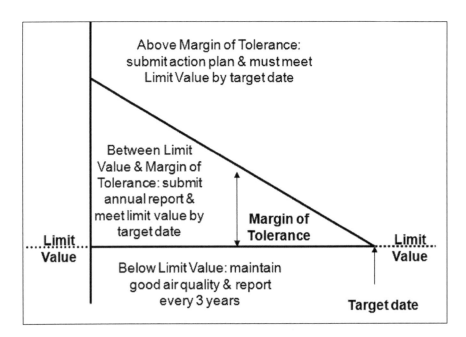

Fig. 9.6 The EU air quality management framework

9.8 Developing Local Air Quality Management Options

As pollution climate can vary markedly over short distances, management may be better delivered at a local level. This is particularly true for cultural heritage as it is often located within towns and cities where pollutant concentrations can be very different from area 'background' concentrations or those calculated using transboundary models. In developing a plan it is important to develop as wide a range of options as possible and score these against agreed criteria including their contribution to sustainable development and environmental justice.

Improvement in local air quality can potentially be delivered through regulation, planning, and encouraging behaviour change. By setting emission limits regulation can be a major driver for the introduction of new and cleaner technology. Planning can play a major role in reducing air pollution through both town planning and building design. Consideration of the air flow around an individual building, or group of buildings can ensure that emissions from boilers or other sources are located so as to minimise ground-level concentrations. Behaviour change can be encouraged through publicity campaigns and financial incentives. In developing options the following are some of the factors that it may be helpful to consider:

- Degree of pollution reduction that will be achieved
- Cost
- Links to other policy objectives
- Possible adverse effects
- Effects on competitiveness
- Environmental justice

These are relevant considerations at both international and local level. The degree of pollutant reduction and costs have already been discussed but it is worth briefly considering the other factors.

In some cases it is possible to deliver multiple objectives through a single initiative. For example, pedestrianisation of a town centre can improve road safety, give better visual amenity and improve health by encouraging walking, as well as reducing pollutant concentrations. But there can be disadvantages, possibly more difficult access for the disabled and short-term reductions in trade for local stores. It is necessary to recognise any adverse effects of lower pollutant levels and accept trade-offs where policy objectives are in conflict. It is important not to assume that the benefits will always outweigh the drawbacks.

In general more disadvantaged sections of society experience a worse local environment and, specifically, are often exposed to higher levels of air pollution (Mitchell and Dorling 2003). In developing policies it is important to consider this and ensure that inequalities are not exacerbated and, if possible, reduced (Gegisian 2007).

Table 9.4 Examples of policy options to reduce NO_x concentrations in a town centre

Option	Non Air quality benefits	Potential disadvantages	Knock on consequences
Increase public transport provision	More effective use of road space Reduces social exclusion for non-car owners	High level of financial commitment for infrastructure development	Increase local taxes to provide funding
Low emission zones	Long-term improvement to urban environment Encourages development of alternative technologies, fuels and mobility modes	Potential for social prejudice against owners of older vehicles Enforcement difficulties and resource cost	Potential for no overall emissions decrease for urban area as a whole Potential displacement of vehicles to other locations
Pedestrian town centre	Reduced risk of accidents	Reduced accessibility	Need to improve public transport access to the area
	Longer-term potential for increased trade and economic activity	Short-term loss of trade	Potential displacement of traffic to elsewhere in the vicinity
	Improved overall town environment with lower noise levels	Limited vehicle access may compromise situation	Increased out of hours commercial deliveries
	Encourages healthier lifestyle	Accessibility provision for the disabled	
	Encourages social interaction		

Table 9.4 summarises guidance (UK Environmental Protection 2007) to assist in the preparation of air quality action plans. But the approaches outlined have wider applicability to any urban area aiming to reduce emissions from motor vehicles whether to protect human health, the built environment or specific cultural heritage objects. The examples shown are:

- Increase public transport
- Low emission zones
- Pedestrian town centres

In each case other benefits and potential disadvantages have been considered, along with potential knock-on effects of implementing the option. These demonstrate examples of linked policy objectives, adverse effects, effects on competitiveness and environmental justice. Examples of good practice have been published (Air Quality Archive 2007).

9.9 Tolerable Levels of Pollution?

Figure 9.7 illustrates differences in dose-response functions for systems with a pollutant threshold (critical level), no pollutant threshold level, and with a response even in the absence of pollution. Effects on cultural heritage fall into the third category and it is necessary to define a tolerable level (K_{tol}) to aid policy formulation. This is the maximum level of exposure/dose for which the associated response is deemed to be tolerable. It recognises that there will be residual adverse effects and that these effects could be further reduced by a reduction in pollution levels, but socio-economic considerations indicate that the funds necessary to do so could be better spent in another way.

A key challenge, therefore, in studying the impacts of air pollution on cultural heritage is determining what this level should be (Kucera and Fitz 1995). The current approach is to define the tolerable level as a multiple of a background level (K_b).

$$K_{tol} = nxK_b$$

Where

K_{tol} is the tolerable corrosion rate,
K_b is the 'background' corrosion rate
n is a factor based on technical & economic factors, including 'lifetime' of
materials.

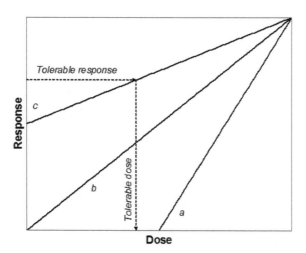

Fig. 9.7 Differences in Dose-Response Functions for systems (a) with a pollutant threshold (Critical Level), (b) no pollutant threshold level and (c) with a response even in the absence of pollution

Within the UNECE ICP on Materials it was decided to recommend the background corrosion or deterioration rate (K_b) as the lower 10 percentile of the observed corrosion rates in the materials exposure programme (UNECE 2004).

Although the definition of a background level based on a percentile of a measured level may be regarded as a procedural convenience, it can be defended on the grounds that there is no implication of a 'safe' endpoint and that such a definition of K_b implies a continuing commitment to pollution reduction. By adopting a K_{90} background level, the target tolerable value falls as pollution levels decrease.

Defining a tolerable corrosion rate draws on experience from restoration and conservation work. In particular to address the questions: 'what is the tolerable corrosion depth before action' and 'what is the tolerable time between maintenance events?' As, even if budgets permitted, it would not be desirable to undertake work too frequently as there are other risks associated with repair and maintenance. In this way average tolerable corrosion rates for cultural heritage materials can be defined for policy purposes, as illustrated in Table 9.5 (MULTI-ASSESS 2007). The table is illustrative and should not be interpreted as a definitive statement.

These data can then be combined with the appropriate background corrosion rates (UNECE 2004) to calculate a tolerable factor, n. Example values are shown in Table 9.6 which are in line with value of 2.5 for n described in Chapter 8.

It is then possible to map areas where current corrosion rates exceed tolerable values and, where data are available, overlay these on maps of cultural heritage to produce basic risk maps. For example, Fig. 9.8 shows areas where current rates of zinc corrosion exceed tolerable levels and areas of high cultural heritage value in Vienna.

Table 9.5 Average tolerable rates for corrosion and maintenance intervals for some cultural heritage materials

Material	Type of surface	Corrosion depth before action (μm)	Tolerable time between maintenance (years)	Tolerable corrosion rate ($\mu m \ yr^{-1}$)
Limestone/ marble	Ornament Aged	100	12	8.3
	Ornament Corroded	50	6	8.3
Calcareous sandstone	Ornament Aged	100	12	8.3
	Ornament Corroded	50	6	8.3
Copper Monument	Ornament Corroded	10	20	0.5
Bronze Monument	Ornament Corroded	10	15	0.7

Table 9.6 Example calculated tolerable factors (n) for a range of materials

Material	Background corrosion rate (μm per year)	Tolerable corrosion rate (μm per year)	Factor, n
Limestone	3.2	8.3	2.6
Calc. Sandstone	2.8	8.3	3.0
Bronze (corr)	0.25	0.7	2.8
Copper (corr)	0.3	0.5	1.7

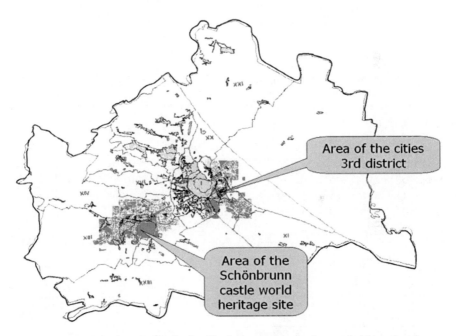

Fig. 9.8 Areas where the calculated tolerable zinc corrosion rate is exceeded (*green*) and areas with a high density of important monuments (*pink*) in Vienna

The derivation of dose-response functions for a multi-pollutant situation has been described in Chapter 3. Based on the dose-response function and tolerable corrosion rate for the material of interest it is possible to calculate a tolerable pollutant situation. This is relatively straightforward for a single pollutant but more complex in a multi-pollutant situation as illustrated in Fig. 9.9.

These relationships form an important element of integrated impact assessments as described later.

Using background corrosion rates and typical 'average' and 'urban' pollution scenarios, SO_2 concentrations for the protection of cultural heritage have been calculated (Multi-Assess 2007). Although drawing on a limited database,

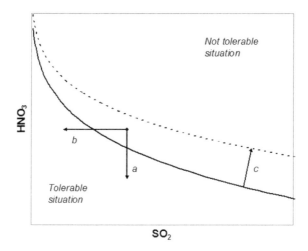

Fig. 9.9 Possible options for changing an intolerable situation to a tolerable one: (a) lowering the HNO₃ concentration, (b) lowering the SO₂ concentration, (c) decreasing the effect of other parameters in the dose-response function such as PM₁₀ or (d) changing the perception of what is considered tolerable

this work has estimated tolerable SO_2 concentrations for a range of materials. The results suggest that a concentration of 10 μg m^{-3} would protect a broad range of heritage materials over 80% of European territory at current HNO₃ levels. This concentration is significantly lower than the annual mean of 20 μg m^{-3} proposed for the protection of vegetation and ecosystems.

Similar calculations have estimated tolerable PM_{10} concentrations for a range of materials and maintenance intervals based on an acceptable 35% loss in reflectance, as shown in Table 9.7.

Table 9.7 Examples of calculated tolerable levels of PM_{10} (μg m^{-3}) for a range of materials and different maintenance intervals

Material	5 years	10 years	20 years
Painted steel	40	20	10
White plastic	45	22	11
Limestone	36	18	9
Average	40	20	10

9.10 How do we Address a Policy Deficit?

In assessing the need for further policy and measures, an initial step is to compare current and future concentrations (predicted on the basis of current policy) with estimated tolerable pollutant concentrations. This will provide an indication of likely future rates of corrosion. If these are greater than deemed acceptable then additional policy measures will be necessary.

Intervention to reduce ambient pollutant concentrations can take place at a range of spatial scales – European, national and local. In each case it is necessary to consider the costs and benefits. If rates of corrosion are severe over a wide area then international effort may be appropriate. If localised, then local air quality management may offer a more cost-effective way forward. In either case pollutant levels, rates of corrosion, costs of damage must be considered alongside the cost of abatement strategies. It is also important to remember the elements of good policy making described earlier.

9.11 International Policy

By considering the collective impacts of air pollution and wet deposition it is possible to undertake analyses of the environmental benefit arising from various emission reduction scenarios. The process is shown schematically in Fig. 9.10

Although there can be considerable uncertainties in this type of analysis, techniques from the field of risk assessment can be used to examine the work systematically and identify those assumptions and uncertainties which could have significant consequences (ApSimon et al., 2002). Following analyses of a range of scenarios within the Clean Air for Europe programme (CAFÉ) the European Commission has presented a thematic strategy for air pollution (EC 2005). This establishes interim environmental air quality objectives for the EU up to 2020. The analyses are summarised in Table 9.8. If implemented the policies should significantly improve air quality and reduce impacts on health and ecosystems. They will also lead to reductions in the rate of deterioration of cultural heritage. One policy instrument to deliver the strategy will be a revision of the Directive on National Emissions Ceilings.

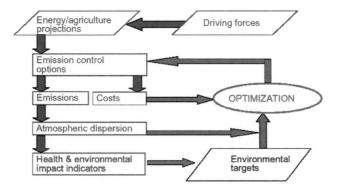

Fig. 9.10 The RAINS model demonstrates the inter-relationship of scientific measurements, economic analyses and policy instruments in establishing optimal air quality levels. (from Amann et al., 2004)

Table 9.8 Alternative environmental scenarios up to 2020

Level of ambition	Benefits		Natural environment (1000 km^2)		
	Human health		Acidification (forested area exceeded)	Eutrophication (ecosystem area exceeded)	Ozone (forest area exceeded)
	Monetised health benefits (€Bn)	Life years lost due to PM$_{2.5}$(million)	Premature deaths due to fine particles and O$_3$		
2000		3.62	370,000	243	733	827
Baseline 2020		2.47	293,000	119	590	764
Thematic strategy 2020	42–135	1.91	230,000	63	416	699
MTFR 2020	56–181	1.72	208,000	36	193	381

(1) CAFE baseline 2020 is the expected evolution of pollutant emissions in the EU-25 up to 2020 assuming that all current legislation to reduce air pollution is implemented. The baseline is based on forecasts of economic growth and changes in energy production, transport and other polluting activities.
(2) MTFR is the Maximum Feasible Technical Reduction and includes the application of all possible technical abatement measures irrespective of cost.

To date these analyses have not taken explicit account of impacts on buildings materials and cultural heritage, but have gone beyond earlier work in describing the effects in detail (Holland and Watkiss 2004). Recent research however, as reported in this volume, will allow improved assessment of the impacts of current and projected air pollution levels on cultural heritage materials that may serve as an input model into the RAINS integrated assessment model and other, more refined, models at regional and city levels.

9.11.1 Role of Standards

At the European level it may be appropriate to introduce air quality standards or limit values. In the case of cultural heritage it is long-term averages, rather than short episodes of elevated concentration, that are likely to be of concern. In developing a limit value or standard there are a number of steps to consider (RCEP 1998).

- Observation of a detrimental effect
- Establish the cause(s)
- Analyse the robustness of the database
- Consider confounding factors
- Consider the spatial spread
- Determine the dose-response relationship

So, while science is an essential starting point it is important to take account of the full range of knowledge and possibilities, indicate where the boundaries of current knowledge lie and make it clear what degree of protection the standard is intended to afford. Standards must draw on rigorous and dispassionate analysis, take account of people's views, be subject to independent peer review and, in the environmental field, sit within a framework of sustainable development. Standards are most appropriate when an effect can be observed over a wide area and where the receptor is widespread, for example human health. When a sensitive receptor is not widespread, perhaps occurring in only a small fraction of the land area, and if pollutant concentrations in those areas may be heavily influenced by comparatively local sources, then other forms of action may be appropriate. In such cases local air quality management may offer a more cost-effective remedy.

9.11.2 Exposure Reduction Approach

In cases where there is a very low or zero threshold it may be more beneficial to ensure an overall reduction in pollutant concentrations rather than focusing on areas of highest concentration, as for $PM_{2.5}$ in the 2008 Air Quality Framework Directive. Figure 9.11 shows a theoretical representation of the relative benefits

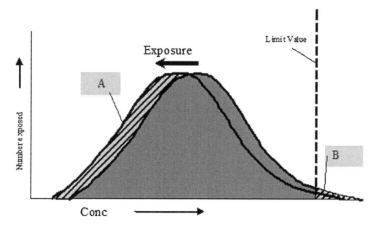

Fig. 9.11 Theoretical exposure reduction approach. (Defra 2007)

of a conventional limit values approach and the exposure reduction approach in the case of human health (Defra 2007). Reducing exposure in hotspots where relatively small numbers of people may be exposed (B) will be less effective than seeking to reduce average population exposure (A). Similar arguments may apply in the case of cultural heritage although there are significant differences in the nature of the spatial distribution.

9.12 Conclusions

Throughout the world our lives are enriched by a wonderful heritage in stone and metal representing the cultures and skills of our ancestors. Today this heritage is under threat – from development on an unprecedented scale, illicit trade, wilful destruction and also air pollution. The problem is well recognised and action is being taken in many parts of the world. One example is the recent Partnership for Sustainable Urban Transport in Asia (PSUTA). This started in 2004, with Xi'an (China), Hanoi (Viet Nam), and Pune (India), to develop a framework through the use of indicators that each city can follow to develop a medium term sustainable transport strategy. The RAPIDC Project described as a case study in Chapter 3 is another excellent example of international coopera-tion in South Asia and Southern Africa to reduce the effects of pollutants.

It is important that the impacts of air pollution on our cultural heritage are now routinely considered alongside effects on human health and ecosystems in developing national and international air quality policy. We have a sound scientific base from which to work, providing information on the levels of pollution that may cause damage and the likely efficacy of actions at local, national and international scales. But conflict can emerge as different groups identify different priorities, which necessarily have to compete for scarce resources. The well known example of India's most famous monument, the Taj Mahal, was an example of the problem. Conservation policy measures conflicted with the wishes of local industry and the workers employed by it. Resolution of air quality issues of this type is required to protect our monuments but at the same time it is important that people have work.

These are complex issues and the risk posed by air pollution is only one threat to our heritage. Time is not on our side but the many advances made through initiatives such as the international charters discussed elsewhere in this volume show that substantial progress can be made. We hope the contents of this book will help national and international authorities frame air quality standards and implement actions so that future generations can continue to enjoy and learn from our heritage as we have done. It is also our hope that some of the insights from science and economics will help all of those who take some responsibility for looking after our heritage. The numerous examples and case studies we have included show that a great deal of work is being done and, hopefully, they will inspire others. It is clear that problems of this complexity need to be resolved by gaining consensus and building on the work of many individuals from many different disciplines. It is our hope that the multi-disciplinary approach to this research has been made clear. We all have a stake in the future of our heritage and we can all make an important contribution. Artists and historians have just as much of a voice in the debate about tolerable levels as do scientists. Hopefully this volume has presented ideas, techniques and concepts from a useful range of associated fields to help everybody move forward. This is important work.

References

Air Quality Archive (2007) UK air quality archive; action planning http://www.airquality.co.uk/archive/laqm/ap_good practice.php

Amann M, Bertrok I, Cofala J, Heyes C, Klimont Z, Schopp W, Winwarter W (2004) Baseline scenarios for the Clean Air for Europe (CAFÉ) programme. Final report to the European Commission http://www.iiasa.ac.at/rains/index.html

ApSimon HM, Warren RF, Kayin S (2002) Addressing uncertainty in environmental modelling: a case study of integrated assessment of strategies to combat long-range transboundary air pollution. Atmos Environ 36:5417–5426.

Baldauf R, McDonald J, Heck R, Cook R, Rege J, Audette L, Armstrong J (2004) Management of motor vehicle emissions in the United States. In: Regional and local aspects of air quality management, Chapter 5. WIT Press, Gateshead.

CCE (2005) CCE status report (Posch M, Slootweg J, Hettelingh JP eds) European Critical Loads and Dynamic Modelling. RIVM report 259101016, Bilthoven http://www.mnp.nl/cce/

COMEAP (2007) Committee on the Medical Effects of Air Pollutants 'Long-term Exposure to Air Pollution: effect on mortality http://www.advisorybodies.doh.gov.uk/statementsreports/longtermeffects2007

CORDIS (2007) Community research and development information service. cordis.europa.eu

Defra (2007) The air quality strategy for England, Scotland, Wales and Northern Ireland (Volume 1). The Stationery Office, London.

EEA (2007a) NO$_x$ emissions from motor vehicles http://dataservice.eea.europa.eu/atlas/view-data/viewpub.asp/id = 360

EEA (2007b) Europe's environment – the fourth assessment. European Environment Agency, Copenhagen http://www.eea.europa.eu/pan-european/fourth-assessment

EEA (2007c) Air pollution in Europe 1990–2004. European Environment Agency 2/2007, Copenhagen http://reports.eea.europa.eu/eea_report_2007_2/en

EC (2005) The communication on thematic strategy on air pollution and the directive on ambient air quality and cleaner air for Europe, impact assessment http://europa.eu.int/com/environment/air/café/pdf/ia_report-en050921_final.pdf

EPA (2007a) The plain English guide to the Clean Air Act. US EPA, Research Triangle Park NC http://www.epa.gov.air/caa

EPA (2007b) National ambient air quality standards. US EPA, Research Triangle Park NC http://www.epa.gov.air/criteria

EU (2007) Directives, environment, air pollution http://europa.eu/scadplus/leg/en/S15004.htm

Europa (2007) European Union air quality standards http://ec.europa.eu/environment/air/quality/standards/htm

Gegisian I (2007) Assessing the contribution of local air quality management to environmental justice in England and Wales. PhD thesis, Faculty of Applied Sciences, University of the West of England, Bristol.

Holland M, Watkiss P (2004) Economic assessment of materials damage and the CAFÉ cost-benefit analysis. In: Cultural heritage in the city of tomorrow (Kucera V, Tidblad J, Hamilton R Eds). Swedish Corrosion Institute, Bulletin 110E, Stockholm.

Irwin JG, Duarte-Davidson R, Pollard SJT (2002) Characteristics of environmental harm in the context of air pollution. In: Brebbia CA and Martin-Duque JF (Eds) Air Pollution X, pp 191-199. WIT Press, Southampton.

Kucera V, Fitz S (1995) Direct and indirect air pollution effects on materials including cultural monuments. Water, Air, and Soil pollution 85:153–165.

Mitchell G, Dorling D (2003) An environmental justice analysis of British air quality. Environment and Planning A 35:909–929

Multi-Assess (2007) Model for multi-pollutant impact and assessment of threshold levels for cultural heritage. EU 5FP RTD project contract: EVK4-CT-2001-00044.

NEGTAP (2001) Transboundary Air Pollution – acidification, eutrophication and ground-level ozone in the UK http://www.nbu.ac.uk/negtap/

Nilsson J, Grennfelt P Eds (1988) Critical loads for sulphur and nitrogen. Environmental report 1988:15, Nordic Council of Ministers, Copenhagen.

UK Environmental Protection (2007) Guidance on local air quality management http://www.environmental-protection.org.uk/assets/library/documents/AQMAprocedures.pdf

RCEP (1998) Royal Commission on Environmental Pollution 21st report, Setting Environmental Standards http://www.defra.gov.uk/environment/rcep/21/index.htm

Stedman JR, Maynard RL (2007) Particles as air pollutants 2: particulate matter concentrations in the United Kingdom. Chemical Hazards and Poisons Research 9:27–33.

UNECE (1999) Gothenburg protocol to abate acidification, eutrophication and ground-level ozone. Protocol to the UNECE Convention on Long-Range Transboundary Air Pollution.

UNECE (2004) Manual on methodologies and criteria for mapping critical load and levels and air pollution effects, risks and trends http://www.oekodata.com/pub/mapping/mapp-man_2004.pdf

Williams ML (2007) UK air quality in 2050 – synergies with climate change policies. Environmental Science Policy 10:169–175.

Wilsdon J, Willis R (2004) See-through Science. Demos, London.

WHO (2006) WHO air quality guidelines for particulate matter, ozone, nitrogen dioxide and sulphur dioxide. Global update 2005, summary of risk assessment. WHO/SDE/PHE/OEH06.02.

Sources of Additional Information

Air pollution climate.
Europe's environment – the fourth assessment.
www.eea.europa.eu/pan-european/fourth-assessment
Brimblecombe P (1987) The Big Smoke. A history of air pollution in London since medieval times. Routledge, 185 pp.
Air pollution in Europe 1990-2004. Copenhagen: EEA 2/2007.
reports.eea.europa.eu/eea_report_2007_2/en
Policy development.
RCEP (1998) Royal Commission on Environmental Pollution 21st report, Setting Environmental Standards.
www.defra.gov.uk/environment/rcep/21/index.htm
European regulatory regime.
EC (2005) The communication on thematic strategy on air pollution and the directive on ambient air quality and cleaner air for Europe, COM(2005)446.
europa.eu.int/com/environment/air/café/pdf/ia_report-en050921_final.pdf
www.unece.org/env/lrtap/
Local air quality management.
UK air quality archive action planning.
www.airquality.co.uk/archive/laqm/laqm.php
NSCA local air quality management.
www.nsca.org.uk/pages/policy_areas/local_air_quality_management.cfm

Index

A

Acceptable levels of pollution, 285–288
Accumulation mode, 19
Acid rain, 1, 56, 59–60, 87, 141
Actinomycetes, 139–140
Aerodynamic diameter, 19
Aesthetic value, 56, 204, 257
Air movements around complex structures, 227–236
  balsa models of gothic or tower roof, 232
  gothic tower, 228
  miniature hole in model roof, 233
  modelled velocity streamlines, 229
  modelling of air flow around towers, 231–233
    experimental mean pressures values *vs.* numerical results, 234, 235
    results and application, 233–236
    scaled validation model of tower, 235
    stone deterioration of black tower, 235
    surfaces with soiling levels, 236
    tube system connecting pressure measurement spots, 233
  pressure taps on benchmark model, 232
  streamlines characterizing flow, 230
  wind tunnel tests, 230
Air pollution, mapping, 44
  UK, 44–47
  London, 47–49
Air pollution, sources, 1–2, 7, 9, 10–11
  indoor, 247–254
  SO_2, 12
  NO_x, 15
  PM, 20–21
  VOCs and PAN, 23
Air Quality Framework Directive (AQFD), 275
  management framework, 282
Air quality policy
  average tolerable rates, 286

differences in dose-response functions, 285
example calculated tolerable factors, 287
options for changing intolerable situation to tolerable one, 288
tolerable levels of PM_{10}, 288
zinc corrosion exceed tolerable levels and high cultural heritage value areas, 286
addressing policy deficit, 288–289
critical loads and levels, 279–281
  critical loads exceedances for acidity, 281
elements of good policy, 272–275
factors in policy development, 270–272
international policy, 289–292
  alternative environmental scenarios up to 2020, 290
  exposure reduction approach, 291–292
  role of standards, 291
  theoretical exposure reduction approach, 292
local air quality management, 281–282
regulatory regime in Europe
  EU limit and target values, 276–277, 278
  EU limit values and UK air quality objectives, 278
  national air quality objectives, 278
regulatory regime in USA, 279
Air Quality Strategy (AQS), 47
AirQuis, 41
Aksu, R., 37, 114
Algae, 134, 136
Ambient Air Quality Framework Directive, 276
Amann, M., 289
ANSI/ISA system, 255

297

ApSimon, H.M., 149, 289
ArcGis software, 155
Arc View GIS System, 152
Atmosphere, mixing layer of, 7
Atmospheric particulate material, 20
    primary/secondary, 21
Atmospheric pollutants, 5–6
Automatic monitoring networks
    in Europe, 32–33
    in USA, 35–36
        NAMS network, 36
        SLAMS network, 35
Automatic Urban and Rural Network (AURN),
    17, 33, 34
Average European conservation and renovation
        costs, 194

B
Bacci, M., 219
Backscattered electrons (BSE), 118
Bailey, D.L.R., 17
Baldauf, R., 281
Beloin, N.J., 109
Biocorrosion, 128
Biodeterioration, 128, 130
Biofilms, 136
Biological agents
assimilation and dissimilation damage, 3
Biological weathering and air pollution
    cost and benefit of biological growth, 142
    gaps of knowledge and future work,
        142–143
    interactions, 140–142
    material destruction process
        physical and chemical transfer
            reactions, 128
    materials and environment, 130–132
        biological pitting destruction, 131
        rock eating fungus, 131
    previous work, 129–130
    resistant rocks, 132
    resistant wood, 132
    types of organisms and deteriorative
            potential
        chemolithotrophic microorganisms,
            137–138
        chemoorganotrophic microorganisms,
            139–140
        effects of trees and roots on buildings or
            tombstones, 134
        iron and manganese oxidizing
            microorganisms, 138
    macroscopically visible organisms, 134
    microorganisms in general, 134
    phototrophic microorganisms, 135–137
    vulnerable rocks, 132
Black carbon/graphitic carbon, 21
Black smoke concentration, 17
    annual mean black smoke concentrations,
        271
Blades, N., 218, 247
Bottom up approach, 8
Boundary Layer Wind Tunnel (BLWT), 231
Brick churches
    identikit for, 168–169
    quantity of stone, brick and glass, 168
Briggs, D.J., 44
Brimblecombe, P., 2, 61, 247, 249
British Standard BS 8221, 260
Bromofluorocarbons (halons), 7
Bull, K., 209
Bullock, L., 219

C
CADW (historic monuments), 259
Camuffo, D., 249
Caneva, G. E., 142
Cano-Ruiz, J.A., 249
Carbonaceous particulate material, 21
Carmichael, G.R., 39
'Case study - CBA of damage to heritage
        materials in Europe'
"Business as Usual" scenario, 210
Cass, R.G., 249
Chaloulakou, A., 249
Chemolithotrophic microorganisms
    nitrifying bacteria, 138
    sulphur compound oxidizers, 137
Chemoorganotrophic microorganisms
    fungi & actinomycetes, 139–140
    organic materials on rock and mineral
            surfaces, 139
Chen, F., 249
Chen, J., 134
Chlorofluorocarbons (CFCs or freons), 6–7
Chock, D.P., 39
City scale stock at risk estimates
    Madrid, 156
        movable and immovable cultural
                heritage distribution, 156
    Milan, 157
        spatial distribution of stone cultural
                heritage with SO_2 pollution, 157
Clayton, P., 17
Clean Air for Europe (CAFE), 208, 289

Clean Air Initiative for Asian Cities
    (CAI-Asia), 292
Coarse particle mode, 19
Cobb, G., 171
Community value, 257
Computational Fluid Dynamics (CFD), 249
Computational Wind Engineering (CWE), 231
Computer controlled scanning electron
    microscopy (CCSEM), 118
Consequences of past stone cleaning
    intervention on future policy and
    resources, 260
Conservation, 239, 256
  considerations for philosophy, 256
  principles, 256
  regime of monuments and building
    maintenance, factors influencing,
    242
Continental or regional stock at risk, 149–150
  UNESCO cultural heritage sites, 150
Contingent valuation method (CVM), 198, 204
Control of air pollutants
  in indoor environment, 254
Convention on Long-range Transboundary Air
    Pollution (CLRTAP), 63
Copper corrosion at NAPAP test sites, 89
CORNET, 82
Corrosion, 53, 129
  average trends in, 74
  of bronze statues in United States
    and copper, 88–89
    dose-response functions, 96–100
    Hiker study, 89–96
  of carbon steel and zinc, 59
  differences in rates, 69
  dose-response functions
    discussion of terms, 75–76
    early, 76
    multi-pollutant situation, 79–81
    SO_2-dominating situation, 76–79
  dynamic effects, 74–75
    degradation of Portland limestone, 74
  involved phases in, 56
  observed levels, 61
  parameters influence, in tropical and
    subtropical, 87
  processes, 54–61
Corrosion attack, 56
  carbon steel and SO_2 concentration, 71, 86
  first year and annual averages of pollutants
    $vs.$ multipollutant exposure, 72
  in ICP Materials, 68
  of unsheltered carbon steel, 73

Corrosion damage, assessment of, 208–209
Corrosion effects in Asia and Africa
  comparison, 86–87
  CORNET test sites, 83, 85, 86
  environmental and corrosion data, 84
  RAPIDC corrosion programme, 82–84
    test sites and exposure conditions, 83
  results of major programmes, 82
Corrosion processes, 54–61
  degradation of materials, 56
  effect of HNO_3 with temperature/relative
    humidity, 58–59
  effect of particulate matter with
    temperature/relative humidity,
    60–61
  effect of precipitation and acid rain, 59–60
  effect of relative humidity and temperature
    for zinc, 58, 60
  effect of SO_2 with NO_2/O_3 and
    temperature/relative humidity,
    57–58
  mass gain of zinc samples, 57
  ratio of HNO_3 and SO_2 effect, 59
corrosion rate and pollutant concentration,
    relationship, 274
Corrosion rates
  cast bronze, 96–97, 100
  for different materials, 67–70
  guiding corrosion values, 67
  steady state in different types of
    atmosphere, 70
Coryneform/nocardioform bacteria, 140
Cost-benefit analysis, 190–193
  'case study - CBA of damage to heritage
    materials in Europe', 208–211
  extended CBA datasheets, 211
    information required to improve
    assessment, 211–212
    outcome of analysis, 212
  for selected built heritage types, 242–246
Cost-benefit ratio, 191
Costerton, J. W. Z., 141
Cost estimates, 190
Cost function, 75
Cramer, S. D., 89
Creighton, N.P., 109
Critical loads and levels, 280
Crusts, 106
CULT-STRAT project, ix, 80–81, 194, 243
Cultural heritage
  restoration of, 205
Cultural heritage goods
  benefits, 202

Current depositing pollutants, case study, 50
Current levels of ambient pollution
    national (UK) scale, case study, 44–47
        ozone, 46
        PM$_{10}$, 47
        SO$_2$ pollution, 45
    urban (London) scale, 47–49
        levels of acid deposition, 49
        ozone, 48
        PM$_{10}$ concentrations, 49
        PM$_{10}$ emission, 48
Cyanobacteria, 135
Cyrys, J., 44
Czop, J., 247, 255

D
Dahlin, E., 219, 220
Damage function, 75
Damages, 75
Data sources
    to identify localities and individual
        buildings, 164
Davidson, C. I., 110
Defra, 278, 292
Degradation, 129
Del Monte, M., 106
Denison, P.J., 37
Department for Environment, Food and Rural
        Affairs (DEFRA), 11
Deposition monitoring
    passive samplers for gaseous pollutants and
        particulate matter, 38
    sampling racks, 36
Deposition rate, 31, 37
Desulfovibrio desulfuricans, 137
Deterioration, forms of
    associated with atmospheric pollution, 2–3
Diakumaku, E. A. A., 142
Dimitropoulou, C., 249
Direct costs for maintenance and repair,
        193–194
Direct Measurement Method, 158, 186
Dorling, D., 283
Dornieden, 139, 143
Dose, 75
Dose-response functions, 30, 37, 54, 75
    corrosion, 75
        D-R functions for SO$_2$ dominating
            siutaion, 76
        D-R functions for multi-pollutant
            situation, 79
    corrosion attack of unsheltered copper *vs.*
        pH of precipitation, 78

        exposure of unsheltered materials, 79
        materials, parameters and inclusion, 79, 80
        soiling, 122
Druzik, J.R., 249
Dry deposition, 58, 66, 78
Dyer, B. D., 128

E
EC Directives, 32
Economic damage function, 75
Economic evaluation
    application of CVM to cultural heritage
        goods, 203–204
    calculation of direct cost for maintenance
        and repair, 193–197
        assumptions, 195–196
        average European conservation and
            renovation costs, 194
        estimation of air pollutant dose–
            exposure cost, 197
        model for calculating costs due to air
            pollution, 196–197
    cost benefit analysis in relation to cultural
        heritage, 191–193
    cost benefit analysis of damage to heritage
        materials
        in Europe, 210–211
        methodology for corrosion effects,
            208–209
        methodology for soiling effects of
            particles, 209–210
    development of datasheets for extended
        CBA, 211–212
    outcome, 212
    effects on local economy, 205–208
        economic impact of tourism, 207
        employment impacts of cultural
            heritage restoration, 208
        leakage, 206–207
        multiplier effect, 206
    methods for valuing welfare loss of
        damages to cultural heritage,
        198–203
        cost-benefit analysis, 199
        cultural heritage interventions, 201–203
"Economic Evaluation of Air Pollution
        Abatement and Damage to
        Buildings including Cultural
        Heritage," 75
Ekberg, L.E., 248
Electron microprobe analysis (EPMA), 119
Emission inventories, 10
Emission reduction methods, 24

Energy dispersive X-ray spectroscopy (EDS), 119
English Heritage, 258
Environment, pollution and effects
    air pollutants
        ground-level ozone, 22–24
        nitrogen oxides and nitric acid, 14–15
        particulate matter, 17–20
        sulphur dioxide, 12–14
    damage to cultural heritage materials
        chemical and biological damage, 3–4
        physical damage, 3
        soiling, 4
    emission inventories, 10–11
    environmental factors
        air and air pollutants, 5–6
        meteorological and climatological factors, 6–7
        processes in damage caused by air pollution, 9
        radiation, 4
        synergy of weather factors, 7–9
        temperature, 4–5
        water, 5
        wind, 6
    pollutant characteristics
        gaseous or particulate, 10
        natural or anthropogenic, 10
        primary or secondary, 10
    trends and scenarios, 24–26
        contribution of road transport and coal combustion, 26
        reduction for pollutants, 26
        US transportation emissions, 25
Environment & Heritage Service, 258–259
Erosion and abrasion, 129
EU air quality management framework, 282
European Cultural Heritage, 149
European Environment Agency (EEA), 32, 40
European Environment Information and Observation Network (Eionet), 32
European Monitoring and Evaluation Programme (EMEP), 11, 38
European Pollutant Emission Register (EPER), 25
European Topic Centre on Air and Climate Change (ETC/ACC), 40
Eutrophication, 280
Evaluation by direct measurement method, stock of materials on Façades, 171–186
    comparison of different materials in façades of buildings and monuments, 175

frontage on Piazza del Popolo and side on Via del Babuinon, 186
    in Paris
        Ile de la Cité with Notre Dame Cathedral, 173
        satellite view of Paris, 172
        Western extremity of Ile de la Cité, 176
    in Sestiere of Dorsoduro in Venice
        Façade of Sant'Agnese Church, 180
        geographical distribution of total surface, 174, 176, 177
        location of studied area, 179
        risks for materials, 182
        southern bank of Canal Grande, 179
        traditional Venetian popular house, 181
    in Via del Babuino in Rome
        columns half painted, 184
        Map of Rome showing Via del Babuino between Piazza di Spagna and Piazza del Popolo, 183
Evidential value, 257
Experimental market techniques, 198
Exponential model, 112
Exposure reduction approach, 277

F
Fan, Y., 249
Ferm, M., 37
Ferrobacillus ferrooxidans, 138
Ferrobacillus sulfoxidans, 138
Fine particle mode, 20
Finstad, A., 42
Fitz, S., 29, 100, 285
Foster, F., 41
Friedlander, S.K., 19
Fuchs, D.R., 219
fuel efficiency, improvement of, 274
Fungi
    deteriorating effects, 139–140

G
Gaseous pollutants, 6, 10
    indoor, 247–254
    SO_2, 12
    NO_x and nitric acid, 14
Gatz, D. F., 99
Gegisian, I., 283
Gehrmann, C. K., 134, 141
Geike, F.R.S., 61
Geographical information systems (GIS), 11, 44, 261
Goodchild, M.F., 44
Google Earth Programme, 158
Gorbushina, A. A., 129, 130, 132, 140, 141

Graedel, T.E., 88
Greater London Authority (GLA), 11, 47
Greenhouse effect, 8
Griffin, P. S., 132
Grøntoft, T., 29, 215, 249, 250
Ground-level ozone (O₃), 22
    effects, 23
Gypsum crusts, 106
Gysels, K., 249

H
Hamilton, R.S., 1, 21, 29, 105, 109, 110, 113,
    114, 122
Harrison, R.M., 14, 21
Hatts, L., 167
Hayes, S.R., 249
Haynie, F.H., 109, 114
Henriksen, J. F., 100
Hiker bronze statues, 89–100
    lifetime precipitation *vs.* maximum pit
        depth, 95
    maximum pit depth *vs.* dedication year, 94
Hill equation (variable slope sigmoid), 115
Hinds, W.C., 19
Hirsch, P., 142
Historic Scotland, 259
    See also Technical advice notes (TANs)
Hoffland, E., 143
Holland, M.R., 209, 291
Horvath, H., 114

I
ICP Materials, 62, 63
Identikits, 150, 165
    building, 166
IMPACT model, 250–253
    evaluation of results, 253–254
    infiltration conditions in mechanically
        ventilated buildings, 252
    materials for indoor surfaces, 252
    web interface, 250
Implied market decisions revealed preferences
    (RP), 198
Indicators
    building identikit, 166
    City of London Churches, 167–168
    Gothic Cathedrals and Convents, 165–167
        example of building identikit, 166
        spatial distribution, 167
Indirect costs, 198–203
Indoor Air Quality (IAQ), 254
Indoor pollutants, sources of, 247
Information needed for mapping stock at risk
    data sources, 163–164

    key areas, considering, 163
    top down and bottom up approach, 163
    two routes followed to identify relevant
        buildings, 164
    United States, 164–165
Ink-streak flora, 135
Innovative Modelling of Museum Pollution
    and Conservation Thresholds
    (IMPACT), 250, 254
Interdepartmental Group on Costs and Benefits
    (IGCB), 210
International Committee on Monuments and
    Sites (ICOMOS) Charter on the
    Built Vernacular Heritage, 258
International Co-operative Programme (ICP),
    54, 279
Ionescu, A., 105, 115, 116, 147
Irwin, J.G., 269, 272
ISO CORRAG program, 62
ISO TC 156, 255
ISO/TC 156, 65
Italian Touring Club Guide (ITCG), 151

J
Jarrett, D., 122
Jens, K., 138
Johansson, L.-G., 57
John, D. M., 135
Jones, A.M., 21
Jones, A.P., 248
Jones, D., 139
Jongmans, A. G., 139

K
Kitson Hiker Statues, 88–100
    corrosion, 94, 95
    distribution, 90
    pit depth measurements *vs.* corrosion, 99
    surface profiles, 93
    verdigris corrosion, 92
Knotkova, D., 62, 82, 100
Kondratyeva, I. A., 132
Krätschmer, 91
Krumbein, W. E., 127, 128, 130, 132, 134, 138,
    139, 141, 142
Krupnick, A., 209
Kucera, V., 53, 141, 223, 269, 285

L
Lanting, R.W., 114
Larsen, R., 218
Lawrence, R.G., 21
Layton, D.W., 249
Lazaridis, M., 248, 250

Lefèvre, R.A., 105, 147
Lemmons, T.J., 109
Lichen, 134, 135, 142
Limit values/air quality standards
  steps, 291
Lins, A., 89
Lipfert, F. W., 37
Liu, De-Ling, 249, 253
Local air quality management, 281–282
  options, 283–284
Lombardo, T., 110, 115
London Air Quality Network (LAQN), 33
London Atmospheric Emissions Inventory
  (LAEI), 11, 47
Long-Range Transboundary Air Pollution
  (LRTAP), 279

M
Machill, S., 130
Magnus, 61
Maintenance, 237–238
  actions, 239
Major exposure programs, 61–73
  corrosion attack of zinc and SO_2
    concentration, 70
  decreasing trends, 73
  ICP Materials test sites, 64–65
  International Standards, 65–67
  ISO 9223 corrosivity categories, 66
  long-term trends, 70–72
  map of test sites, 63
  pollution parameters, 62
  rates for different materials, 67–70
  recent trends, 72
  UNECE ICP materials programme, 62–65
Mansfield, T.A., 21, 109, 110, 113, 114
Mapping, 44
  mapping corrosion rates of carbon steel, 80
Mass balance models, 249
MASTER project, 220, 253
Matt, D.R., 37
Matthews, G., 263
Mavroidis, I., 249
Maynard, R.L., 271
McMahon, T.A., 37
Mean time to failure (MTTF), 237
Memory effect, 74
  for natural stone buildings, 75
Metallogenium symbioticum, 138
Methane/non-methane VOCs (NMVOC)
  NMVOC emissions, 23
  sources, 23
MICAT project, 62

Micro-erosion meter, 74
Microorganisms, 134
Middleton, P., 167
Mikhailov, A.A., 66
Millipore Isopore Filters, 119
Mitchell, G., 283
Model calculations and scenario modelling of
    urban levels of particulate matter
    and risk of soiling
  cleaning interval in years, 223
  Oslo Reference 2010, 224, 225, 226
Model calculations to estimate urban levels of
    particulate matter, 41–43
  average concentrations of PM_{10}, 42, 43
Model Documentation System (MDS), 40
Modelling, 39–40
Monitoring
  case study, 33–35
  deposition, 36–39
  increase of automatic monitoring stations,
    30
  networks for ambient air, 31–35
    measurement techniques for automatic
      analysers, 32
Monitoring network
  automatic ambient air, 31
Monuments, 148
Morcillo, M., 62, 82
Moveable heritage, threats to, 263
Muller, C., 255
MULTI-ASSESS programme, 62, 79, 110,
    112, 117–118, 121, 154–155, 220,
    223, 286–287
Multiplier effect, 206
Multiplier values, 205

N
National Acid Precipitation Assessment
    Programme (NAPAP), 271
National Air Monitoring Stations (NAMS), 35
National and regional organizations and
    guidance
  legislative guidance
    Czech Republic, 260–262
    published guidance in UK, 260
    Spain, 262–263
  legislative organisations, 258–259
  threats to cultural heritage, 263
National Atmospheric Emissions Inventory
    (NAEI), 11
National Heritage Institute, 261
National Historic Preservation Act of 1966,
    164

National Register of Historic Places
    documentation, 165
National scale stock at risk
    Czech Republic
        limestone corrosion map, 155
    estimating risk using census data, 151
    Italy, 153
        distribution of cultural monuments, 154
    Norway, 154–155
        limestone corrosion map, 154
NATO/CCMS Pilot Study, 61
Navrud, S., 189, 198, 201, 203
Nazaroff, W.W., 249, 253
Newby, P.T., 209
New/revised policy, drivers for, 270
Nicholson, K.W., 37
Nitrogen oxides and nitric acid, 14–16
    annual UK NO$_x$ emissions, 15
    nitrogen dioxide emissions, 16
    sources of atmospheric NO$_x$, 15–16
"Noah's Ark Project", 8
Non-automatic networks, 34–35
Non-market techniques
    categories, 203
Non-market valuation, 203
Norwegian Institute for Air Research (NILU),
    63–64
NO$_x$ emissions from vehicles, 273
NO$_x$ sources, 14, 15
Nuckols, J.R., 44
Nucleation particle mode, 19

O
Oddy, W.A., 219
Oslo Sulphur Dioxide protocol, 149
Owczarek, M., 129
Ozone (O$_3$), 22
    effects, 23
    hole, 7

P
Papakonstantinou, K.A., 249
Parker, A., 108
Particle deposition rate, 114
Particulate elemental carbon (PEC), 21, 114
Particulate matter, 17
    annual PM$_{10}$ emissions, 18
    composition, 21–22
    sources and composition atmospheric,
        20–21
    terminal settling velocities, 19
Partnership for Sustainable Urban Transport in
    Asia (PSUTA), 292
Patnaik, P., 31

Peroxyacetyl nitrate (PAN), 23
Pesava, P., 114
Petersen, K., 132
Petersen, W.B., 41
Photochemical Assessment Monitoring Station
    (PAMS), 36
Phototrophic microorganisms
    biofilm cover, 136
    black biofilm in restored backside of
        Erechtheion, 135
    lichen on Jewish tombstone, 137
    Wall of Neue Pinakothek shows
        'Tintenstrich-Flora (biofilms), 135
Physical damage function, 75
Pio, C.A., 110
Policy options to reduce NO$_x$ concentrations,
    284
Pollutants
    limit values for, 275
    tolerable concentrations of, 255
Pollutants, sources of, 11
Power, T., 89
Prillinger, H.-J., 130

Q
"Quantification of Effects of Air Pollutants on
    Materials", 62

R
Rabl, A., 209, 210
RAINS model, 289
Rapid tool kit, 220
    basic kit, 220–221
    complete kit, 221
    rack, 221
REACH programme, 195, 196–197
Ready, R.C., 198, 203
Reeve, R.N., 31
Regional Air Pollution in Developing
    Countries (RAPIDC), 82
Reilly, J. M., 91
Renovation, 239
Restoration investment
    effects of, 205
Return period (RP), 237
Revealed preference techniques, 203
Risk assessment and management strategies at
    local level
    air movements around complex structures,
        227–236
    conservation & maintenance strategies
        general principles, 256–258
    cost/benefit analyses for built heritage
        types, 242–246

case studies, 243–246
  environmental, economy and life cycle
    data, 243–245
damage caused by air pollution to indoor
    heritage materials, 247–254
  exemplifying models – IMPACT model,
    250–253
  modelling of indoor air quality and
    pollutants infiltrated, 248–250
life time estimates as base for management
    strategies, 236–241
  comparison of costs and life times, 240
  return periods of typical restoration and
    maintenance works, 238
mapping and modelling pollution and risk
    of corrosion and soiling, 222–226
  model calculations and scenario
    modelling, 222–226
monitoring, 217–220
  dosimetry, 219–220
  parameter, 218
in situ evaluation of effects, 220–222
  rack for rapid tool kit, 221
strategies to mitigate outdoor pollution
    effects on indoor environments
  control of air pollutants, 254
  tolerable levels for air pollutants and
    material damage, 255
Risk Map of Cultural Heritage in Italy, 153
Robinson, R., 31
Royal Commission on the Ancient & Historical
    Monuments
  of Scotland, 259
  of Wales, 259
Rural pollutant, 23
Rust Belt, 91

S

Saiz-Jimenez, C., 2, 130
Salmon, L.G., 249
Satellite data, 150
Saunders, D., 219
Schikorr, G., 61, 76
Schikorr, I., 61, 76
Sebera, D. K., 220
Secondary electrons, 118
Sherman, C.A., 41
Sherwood, S. I., 53, 88
Shields, H.C., 248
Slørdal, L.H., 41
Smog, 7
Soiling, 4, 105, 107
  acceptable level of, 124

basic exponential relationship, 114
on buildings, 106–107
dose-response functions
  k vs. PM_{10} and dose-response constant,
    121
  loss in reflectance and soiling constant,
    121
  map for white painted steel, 123
  relationship between soiling constant
    and PM_{10}, 121
  use of, 122–124
  variation in soiling with time/ PM_{10}
    concentration, 122, 124
of glass, 110, 116
levels and trends
  spatial trends and variations, 112
  temporal trends and variations, 108–111
mechanisms and models, 112–117
  basic exponential model, 114
in road tunnel, 109
temporal trends, 110–111
variability in particle assemblage, 117–120
  large particle size, 120
  relationship between mass and
    reflectance loss, 117–118
1-year soiling as % loss of reflectance, 112
Special Purpose Monitoring Stations (SPMS),
    35
Spence, J.W., 109
State and Local Air Monitoring Stations
    (SLAMS), 35
Stated preference techniques, 203
State Historic Preservation Offices (SHPO),
    165
State Implementation Plans (SIPs), 281
State of the Environment Reporting
    Information System (SERIS), 32
Stedman, J.R., 271
Sterflinger, K., 130
Stock at risk, 147
  Aldobrandini Village House, 158–161
    Roman windows in Villa, types of, 160
    Villa Aldobrandini, 160
    Villa Aldobrandini with façade areas,
      159
  different scales
    brick churches, 168–169
    city, 155–157
    continental or regional, 149–150
    district level, 157–158
    indicators, 165–168
    individual buildings, 158–162

Stock at risk (*cont.*)
    information needed for mapping,
      162–165
    national, 150–153
    stone churches, 169–171
    evaluation by direct measurement method,
      171–186
    Santa Maria della Vittoria Church, 161–162
      different parts of frontal façade, 162
      frontal and side facades, 161
Stöckle, B., 91
Stock of buildings
    methods for estimating, 150
    spatial distribution, 150
Stone churches
    estimating total stock at risk of materials,
      171
    identikit for, 170
    quantity of stone, brick and glass, 169–170
Strategic Stone Study, 162, 163
Stratosphere, 6
Streptomyces, 140
Subaerial biofilms, 129, 141
Sulphur dioxide
    annual emissions, 12
    SOx emissions, 13
Surrogate market methods, 198
Swedish International Development
      Cooperation Agency (SIDA), 82
Synergistic effects, 57
Synergy, 57

T
Taylor, J., 218
Technical advice notes (TANs), 260
Temperature inversion, 7
Tetreault, J., 218, 247
Thatcher, T.L., 249
Thermal damage, 3
Thiobacillus ferrooxidans, 138
Thiobacillus sp., 137
Thornbush, M., 124
Tidblad, J., 53, 82, 269
Top down approach, 8
    See also Bottom up approach
Toprak, S., 114
Total Economic Value (DTEV), 200
Total Suspended Particulate matter (TSP), 17,
    20, 114
Tourism
    economic impact of, 207

    leakage
      employment impact, 208
      examples of, 207

U
UNESCO World Heritage List, 158
United Nations Economic Commission for
      Europe (UNECE), 63, 149, 271
United States Environmental Protection
      Agency (USEPA), 11, 20
Urzì, C., 130
U.S. Clean Air Act, 281
US Environmental Protection Agency
      (USEPA), 35

V
Value, 257
    of goods, 201
Valuing Cultural Heritage, 203
Vawda, Y., 39
Vertical trends, 108
Viles, H., 124, 130
Vine, M.F., 44
Volatile Organic Compounds (VOCs), 23, 254

W
Watkiss, P., 291
Watt, J., 105, 118, 122, 147, 189
Welfare loss of damages to cultural heritage,
    198
    categories, 200
Weschler, S. J., 248
Wet deposition of pollutants, 76
Williams, M.L., 272
Willingness to avoid (WTA), 198
Willingness to pay (WTP), 198
Willis, R., 272
Wilsdon, J., 272
Wilson, M. J., 139
Wind effects, 227
World Heritage Convention, 148

X
X-rays, 118

Y
Yates, T., 100, 165, 189, 215

Z
Zallmanzig, J., 61
Zinn, E., 218